Technology of Our Times:

People and Innovation in Optics and Optoelectronics

Editor
Frederick Su

SPIE OPTICAL ENGINEERING PRESS

A Publication of SPIE—The International Society for Optical Engineering
Bellingham, Washington USA

Library of Congress Cataloging-in-Publication Data

Technology of our times : people and innovation in optics and optoelectronics / editor, Frederick Su.
     p.   cm.
    "A publication of SPIE--The International Society for Optical Engi-
neering."
    ISBN 0-8194-0472-1 : $25.00
    1. Optics--History. 2. Optoelectronics--History. 3. Physicists-
-Interviews. 4. Engineers--Interviews. I. Su, Frederick, 1947-
QC352.T43 1990
621.36--dc20
                                                      90-39927
                                                          CIP

Published by
SPIE—The International Society for Optical Engineering
P.O. Box 10
Bellingham, Washington 98227-0010 USA

SPIE gratefully acknowledges the following publishers for permission to
reprint articles included in this publication:
    Institute of Electrical and Electronic Engineers; Astronomical Society
of the Pacific; Chuokoron-sha, Inc.; Optical Society of America; *The Retired
Officer* magazine.

Printed in the United States of America

Table of Contents

Education

Cover photographs
Large photo: Nova Laser amplifier chains. (University of California, Lawrence Livermore National Laboratory, and U.S. Department of Energy.) *Upper left:* Natural flexure modes of 8-meter thin meniscus showing surface deformation. (European Southern Observatory.) *Lower right:* Back Dive, (© 1964 Harold Edgerton. Courtesy Palm Press, Inc.)

Foreword

History is not necessarily ancient history, and there have been numerous optical discoveries and inventions during the past 60 years that compare well with the great achievements of Galileo, Newton, and Thomas Young. Indeed, as one gets older, the how and why of scientific innovations become a matter of increasing interest, making this and other books on the history of science more and more fascinating as the years go by.

When we first came to this country in 1929, optics fell into three well-established branches, namely, geometrical, physical, and physiological. Geometrical optics was mainly concerned with the design of lenses and optical instruments; physical optics covered interference, diffraction, and polarization; while physiological optics included vision and colorimetry. Anything outside the visual region, in the UV or IR, was considered a part of spectroscopy, then a branch of physics. We were not quite sure where to put subjects like optometry, atmospheric phenomena, photometry, fluorescence, or photography, which fell outside the classical subdivisions of optics.

Soon after we arrived, the stock market crashed, and America suffered from the great depression that lasted until World War II. During that period, electron optics and photocells appeared in the early 1930s; lens coating was introduced in 1936; and television came soon after. Color and sound were introduced into professional motion pictures about 1930, but the shortage of money caused many other developments to wait until the war was over.

A glance at the cumulative indexes of the various optical journals provides a good indication of the range of new developments that followed the war. Holography came first, to be quickly followed by the whole range of solid-state physics that began to invade optics in the late 1940s. Infrared detectors soon arrived, and the most important invention of all, namely, the laser in 1960. Who, in 1929, would have guessed that strange compounds such as gallium arsenide, cadmium telluride, or indium antimonide could possibly have optical significance? Indeed, in 1929, subjects like quantum optics, optical computing, image processing, acousto-optics, and fiber communications would have seemed like science fiction.

I still marvel at the way bar codes are printed and scanned at the local supermarkets. The shift from black-and-white to color printing in magazines and newspaper advertising is a little-noticed development that has required a great deal of careful thought and a vast investment of time and money.

It is a wonderful time to be living, a time of great innovation— as the articles in this book will attest. I sometimes wonder, if I had my life to live over again, which of the many optical fields I would try to enter. Any of them would lead to a lifetime of rich and rewarding research and activity.

Rochester, New York   *Rudolf Kingslake*
June 1990

Preface

Passion. For some people, it dictates their existence. A passion for knowledge, for discovery, for excellence. A passion for nature—not nature in the natural history sense, though that is no less noble a pursuit—but passion for nature's underlying secrets. Such passion provides the key that will unlock for the world the development of the technology of our times.

Many who are not scientists or engineers cannot fathom the depths of this passion, nor understand the creativity involved in the physical sciences and engineering. The fiction writer creates his world in his mind and commits it to paper. The painter visualizes and captures a part of his world on canvas. The artist is constrained only by his imagination, tempered by his own judgment and perception of reality.

A scientist or engineer, on the other hand, does not have the latitude of an artist. He or she is constrained by the laws of nature. Therefore the winged horse of the artist's imagination must be reined in to reflect the grounded horse of the scientist/engineer's own experience, because experience is the final arbiter of the nature of reality—whatever the philosophical or artistic discourse about the reality of nature. To explain nature, then, is the domain of science; to translate that knowledge into technology is the domain of engineering.

But for all that, there is no less creativity involved in the sciences and engineering than there is in art. The solutions to nature's complexities involve a process of creative imagination and deductive reasoning constrained by real experience. With this process, scientists can unravel the secrets of nature that will lead to the forming of new technology. And while nature may seem guileless at first, as our technology advances, so too, do we discover her underlying complexities. She is, to redirect Winston Churchill's famous quote, "a riddle wrapped in a mystery inside an enigma," like the matryoshka—the Russian nesting doll. If you take apart the outer doll, there is another one inside, and one inside that, and so on.

Technology has transformed our lives, made our lives easier, longer, more productive, and has fed on itself to increase our knowledge of our world and universe. While technology has introduced problems (such as nuclear weapons), it has also made the world better. Who could have foreseen the impact the laser has made? Or the CCD, the CAT scan, xerography, or fiber optics? Who can guess where the new discoveries in chaos, high temperature superconductivity, and other fields will lead us? No one knows. We can only hazard a guess. But there is one concern everyone voices in these highly competitive technical times. Education must play a significant role. If we can spark an interest in the sciences in today's children, and especially encourage girls to become many of the scientists and engineers of tomorrow, then technology—and the United States' role in it—will continue to grow as it has.

Within these pages, then, look for the underlying passion for knowledge and excellence that these researchers have shown. Their discoveries and innovations have peeled a layer from the matryoshka. You will find some of the best articles and interviews from *OE Reports*, SPIE's monthly newspaper. And there are interviews and articles never before published by SPIE, gleaned from other sources or done expressly for this book. They are my choices (yes, there were others, but time and space limitations precluded an exhaustive review) for both a retrospective and a look forward into the technology of our times in optics and opto-electronics.

Acknowledgments

No book is the sole offspring of one parent. Rather, a book like this—a collection of articles and interviews—is a child of an extended family, all of whom have contributed in some manner to its birth. I have tried to provide the guidance and, if I have erred in content, then the fault is mine.

Thanks go to all the contributors—either as authors or as interviewees. Without them, this book would not exist. I would also like to thank Emil Wolf and Brian Thompson for their time and effort, Rudolf Kingslake for graciously writing the Foreword, and John DeVelis for the Afterword.

I would especially like to thank the following people, worldwide: Mollie West, Ed Eckert, AT&T; Rowan Dordick, Martin Hug, June Namioka, IBM; William Herrman, Hughes Aircraft Company; Bill Hagen, IEEE; Andrew Fraknoi, Astronomical Society of the Pacific; Robert B. Smith, Honeywell; Takashi Hirabayashi, Chuokoron; and Robert Reim, St. Joseph Hospital, Bellingham. They smoothed the path by providing contacts, permissions, photos, and line drawings.

Bellingham, Washington
June 1990

Frederick Su
Editor

People and History

We are all living in the gutter, but some of us are looking at the stars.
Oscar Wilde, *Lady Windermere's Fan*

D. J. Lovell

The Legacy of Fourier

Left: Joseph Fourier (The Bettmann Archive, Inc.)
Above: D. J. Lovell

D. J. Lovell, whose long career in optics included periods at MIT, Norwich University, the Naval Research Laboratory, and Los Alamos Scientific Laboratory, was the author of Optical Anecdotes, *published by SPIE in 1981. The popular collection of profiles of noted "optikers" has been reprinted several times. This article appeared as the 20th anecdote in the book.*

In his eulogy of Joseph Fourier to the Paris Academy of Sciences, François Arago concluded, "My object will have been completely attained if . . . each of you has learned that the progress of general physics, of terrestrial physics, and of geology will daily multiply the fertile applications of *Théorie de la Chaleur,* and that this work will transmit the name of Fourier down to the remotest posterity." Although Arago thus predicted a legacy of the use of Fourier's mathematical treatment used to describe the conduction

Reprinted from *Optical Anecdotes,* D.J. Lovell, SPIE, 1981.

of heat, in 1833 he could hardly have envisaged those benefits to be derived in optics. Today, Fourier optics and Fourier transform spectroscopy are widely practiced by scientists with little knowledge of the life of Joseph Fourier.

It is of interest to review Fourier's life and to reflect on the manner of man who brought to us a keener understanding of optical phenomena. Much of my information about his life is derived from the aforementioned eulogy, which was written by a friend and admirer. It was also written after a turbulent period in French history when the passion of patriotism gave rest to overindulgent attitudes toward behavior. I cite this to indicate that I have tried to separate fact from fancy, and to warn you that my attempts to present a clear account of Fourier's life may, at times, be colored by the biases introduced into the documents available to me.

Fourier was born at Auxerre, about 160 kilometers southeast of Paris, on March 21, 1768; he was orphaned at the age of eight. A neighbor lady, who recognized his courteous manners and his precocious natural abilities, recommended him to the Bishop of Auxerre, and Fourier was admitted to the military school run by the local Benedictines. His precocity was soon evident, as Fourier anonymously authored many of the sermons delivered by high dignitaries of churches in Paris.

Although evidently a gifted child, Fourier was also petulant, noisy, and vivacious. However, during his fourteenth year he became interested in mathematics and settled down (or as Arago writes, "He became sensible of his real vocation.").

Educated in a military school directed by monks, Fourier wavered between a career in the church or the military. He preferred the latter, but this was not then possible because Fourier's father had been a tailor and not of the nobility. Fourier thus entered an abbey, but before taking his vows the social upheaval in France attracted him to a teaching position. He was appointed to the principal chair of mathematics in the Military School of Auxerre.

He soon displayed an unusual talent for lecturing on rhetoric, history, and philosophy, substituting for his colleagues when they became ill. His lectures on these several topics attracted a delighted audience of diverse backgrounds. This characteristic distinguished Fourier throughout his career.

In 1789 Fourier read a paper to the Paris Academy of Sciences on the resolution of numerical equations of all degrees. This work, enunciated by Fourier at the age of twenty-one, formed the cornerstone upon which he developed his future mathematical work. In this paper, Fourier extended some of the prior contributions of Lagrange. Nevertheless, Fourier's accomplishment had little appeal to the pure mathematicians, who felt that it lacked rigor. To the physical scientist, however, Fourier's results were received warmly (and still are) since they simplified calculations.

Returning to Auxerre, Fourier enthusiastically embraced the principles of the Revolution. This was a period of revolution not only politically but in the arts and sciences as well. For example, the reformation of weights and measures was begun at this time, leading to the introduction of the metric system. Alas, the times had taken many of the savants into military activity or, as with Lavoisier, had removed them permanently. (Lavoisier was guillotined).

Fortunately, Napoleon in his rise to power realized the futility of ignorance in building a meaningful empire and he encouraged the creation of schools. In 1794, the Ecole Normale was started and Fourier was rewarded for his patriotism in Auxerre by being appointed to the chair of mathematics. This school lasted but a few months, at which time the Ecole Polytechnique was established. Again, Fourier was called and he responded by embellishing his reputation for clearness, method, and erudition. His lectures attracted a fastidious and wide audience.

Soon after the formation of the Ecole Polytechnique, Napoleon began to dream of restoring Egypt to its ancient splendor. And what better way to accomplish this than by introducing French culture to this now-backward country? Napoleon realized that to achieve this goal he would need more than a mere army. He would require leading scientists, and he chose Gaspard Monge and Claude Louis Berthollet. Both of these men were members of the Paris Academy of Sciences, on the faculty of the Ecole Polytechnique, and recognized as being among the leading scientists of the time. They in turn asked Fourier to join them and he did so. In Cairo they established the Institute of Egypt with Monge as the first president and Fourier as perpetual secretary.

In Egypt, Fourier distinguished himself by extending his mathematical researches to general solutions of algebraic equations, methods of elimination, and indeterminate analysis. His breadth of interest was also manifested by his contributions in general mechanics. He designed an aqueduct to conduct water from the Nile to the Castle of Cairo, espoused a proposal to explore the site of ancient Memphis, gave a descriptive account of the revolutions and manners of Egypt, and designed a wind machine to promote irrigation.

However, Napoleon's dream to rescue the Egyptians from the galling yoke under which they had groaned for ages . . . [and] to bestow upon them without delay all the benefits of European civilization (I quote from Arago's eulogy) was a failure. The uncivilized Egyptians failed to respond to the cultural feast proffered. Napoleon surreptitiously returned to France, as did Monge, leaving Fourier to cope. During Fourier's continued stay in Egypt, Napoleon conquered much of Europe and became the virtual ruler of France. But his fortunes in Egypt never proved successful and within three years of Napoleon's stealthy departure, Fourier and the others went back to France.

Upon his return to France, Fourier was named Prefect of the Department of l'Isère. Although this area was a hotbed of political dissension, Fourier, with great diplomatic skill, soon established harmony among the near-warring factions. The situation was brought to such a quiet state that Fourier could continue his efforts in mathematics and letters. From Grenoble, the principal city of Isère, Fourier wrote his *Théorie Mathématique de la Chaleur*. This was Fourier's outstanding scientific achievement.

Fourier's effort received a mixed reception. The pure mathematicians again pointed out the lack of rigor in his treatment. Pure mathematicians and mathematical physicists have nearly always been at odds, the former disdaining any treatment that avoids the scrutiny of rigid proof and the latter pleased to have a procedure to express the results of their observations.

Fourier recognized that any function whose graph displays a periodicity can be considered to be a sum of sinusoidal functions. That is

$$f(x) = \sum_n A_n \sin nx \ .$$

(The purist may note that I have taken some liberties in expressing the Fourier series. Pshaw.)

This series is now known as a Fourier series. Its real value to optics, of course, is that it leads to an integral transform whereby a periodic function of space, for example, may be transformed to a periodic function of time. This means that the spectral characteristics of a radiating source may be separated by a Michelson interferometer to provide a time-varying signal in which time is not directly related to wavelength. Generally, the transformation is undertaken with a computer although, before computers, other means were employed.

Fourier had submitted his treatment of the conduction of heat, in which his series was fundamental, to the Paris Academy in 1811, for which he was awarded its mathematical prize in 1812. As noted, some reservations were expressed with the favorable judgment. However, Fourier never admitted the validity of this dissension, giving unmistakable evidence near the close of his life that he thought it still unjust by causing this memoir to be published in the Academy records without changing a single word!

This work gave a tremendous impetus to the research of his colleagues who considered the geological heat content, the temperature of celestial regions, and the effects of heat on biological growth. During this period, Napoleon's influence had blossomed and faded. In 1815, Napoleon escaped from Elba and made a triumphal march on Paris. Fourier had mixed reactions to this news. He left Grenoble for Lyons, where some of the royalty had assembled. They greeted Fourier coldly and doubted that Napoleon could have captured nearby Grenoble. Consequently, Fourier was told to return and protect the (already fallen) city. Fourier had barely left Lyons when he was arrested by Hussars and conducted to Napoleon's headquarters. Fourier explained that his duty compelled him to act as he had. Napoleon forgave Fourier, but did not endear himself when he told Fourier, "I have made you what you are."

Fourier was appointed Prefect of the Rhône and given the title of Count—promotions Fourier dared not reject. However, this appointment as Prefect lasted but a short time. Fourier returned to Paris with no income and no financial reserve. It was a turbulent time for many. Napoleon's career ended at Waterloo, and the Bourbons were restored to power in Paris.

Fourier applied for a federal pension for his 15 years of service to his country. He was rudely denied. However, a former student at the Ecole Polytechnique, on learning of Fourier's plight, enabled him to receive the directorship of the Bureau de la Statistique of the Seine.

The Academy of Sciences sought at its first opportunity to elect Fourier to the society. Political intrigue, sanctioned by Louis XVII, prevented anyone who had been associated with Napoleon from election to the Academy. (Arago noted, "In our country, the reign of absurdity does not last long.") A year later, in 1817, the Academy again unanimously nominated Fourier to a place in the section of physics. This time there was royal confirmation without difficulty.

Fourier was now able to spend the last years of his life in retirement and in the discharge of academic duties. He became eloquent in discoursing on those facets of life which he had experienced. There are those who find this type of eloquence somewhat boorish, rather than fascinating. They cite an incident in Fourier's later years as a case in point. Fourier was seated at a table together with some who were strangers to him. One, in particular, was identified as an older officer. To him, Fourier described in great detail the events of a battle that had taken place in Egypt, of which Fourier had some first-hand knowledge. Fourier concluded his recitation of the details of the battle by noting, complacently, that he felt his memory had served him correctly in recalling these events. His companion, who seemed to have been enthralled by this discourse, assured Fourier that his statements were accurate and based this judgment on the fact that he, too, had personal knowledge of the battle, having been head of the Grenadiers involved!

Although endowed with a sturdy constitution, Fourier had adopted the habit of wearing too much clothing. Thus, although he gave the appearance of corpulence he was, in fact, a quite slender man. He abided in a sterile, ovenlike environment, even keeping his windows closed in the heat of summer. Visitors found this to be annoying. As a result of this, Fourier developed an aneurism of the heart. In the spring of 1830 he sustained a fall while descending some stairs. This aggravated his condition and, within two weeks, he died.

Fourier's name is now used as an adjective describing an elegant method of handling several optical processes. Thus, we who work in the field of optics revere this man. He was undoubtedly unaware that the legacy he left behind was rich beyond all expectations.

T. E. Allibone

Dennis Gabor: The Father of Holography

A tribute to Dennis Gabor—scientist, engineer, inventor, and Nobel Prize winner—was presented by SPIE in 1983 at its International Technical Conference/Europe held in Geneva. The program honoring the inventor of holography included a retrospective of Gabor's life and work. Among the presentations was this biographical sketch by T. E. Allibone, Professor Emeritus, Electrical Engineering, Leeds University, England.

It is a great honor to be asked to deliver this brief account of the life of a very dear friend, an engineer of exceptional merit, the inventor of holography, a Nobel laureate in physics, and a humanist of rare distinction.

Gabor was born in Budapest, Hungary, 1900, the oldest of three boys, children of talented and artistic parents. Father Gabor, a successful engineer, married a woman of great intelligence, and they brought up their family in the best liberal traditions, employing governesses of English, then French, and then German extraction so that the boys were multilingual. The home had a well-stocked library and was the meeting point of many intelligent friends of the parents; the boys mixed with them and shared in their conversation, and thus absorbed a catholic education in the liberal arts and sciences. Dennis said that he and his brothers were familiar with the great paintings of the Masters from the beautiful art books in the home long before they ever saw them in the national galleries of the world. And, although Dennis was not a trained singer, he had a good voice, and throughout his life could sing parts from many of the grand operas in their own languages.

Gabor took the entrance examination for the Technical University of Budapest in 1918, but then had to join the army and eventually served on the Italian front. Here he learned his fourth foreign language and loved the country he saw; ever after that he was fascinated by Italy, ultimately choosing it for retirement.

He returned to Budapest, studying mechanical engineering for two years and then going to the famous Technisches Hochschule in Berlin to study electrical engineering. This double training partly explains his great versatility; he could design the most complicated

Reprinted from *Optical Engineering*, July/August 1983.

physical and electrical apparatus and always had a drawing board in his office. His favorite subject was optics; indeed, he had learned so much in school and at home that he reckoned he had mastered the subject to a far higher standard than was ever taught in college.

In 1924 Dennis took his diploma and decided to stay on to do some research for the higher degree. At that time Steinmetz's theory of electrical transients and traveling waves was much in evidence, but to record transients of a few microseconds, such as are caused by lightning, an oscillograph having a very fast writing speed was essential. In Aachen, Professor Rogowski was working on such an oscillograph, and Dennis began to make one also. He accelerated an electron beam to 60 kV before it entered the deflecting chamber, and his great contribution was to focus the beam with a short solenoid, which he shrouded in iron to concentrate the magnetic lines of force. This was the first electron lens ever to be made in this manner, and it was his first invention. Then he devised a clever electrical circuit, a flip-flop circuit that deflected the beam away from the photographic plate until the electrical transient arrived; since that time, all oscillographs used for recording fast transients have used these two inventions. When the oscillograph was working satisfactorily, Gabor was employed by the Berlin Electric Supply Company to study transients placed on its 100 kV transmission lines, the first investigation of traveling waves on an actual power line.

It was this splendid work that took me to Berlin in 1928 to meet Gabor. I was the first British scientist he had met, and in the afternoon he took me to hear Einstein lecture—an unforgettable occasion. We corresponded on oscillographic matters for the next few years.

In 1927 Dennis was employed by Siemens in Berlin studying a plasma lamp he had invented, but just then, a German mathematician worked out the theory of electron beam focusing, which showed that a magnetic lens behaved exactly like an optical lens with the same laws relating the object distance to the image distance and the focal length; it was immediately realized that an electron microscope might be made. Dennis told the story of how he and Leo Szilard sat in a cafe in Berlin discussing the concept of an electron microscope, but they agreed that it would serve no useful purpose—you could not put living matter into a vacuum, and everything would be burned to a cinder by an electron beam, of say, 100 kV energy. Who would have dared to believe that the cinder would preserve not only the structure of the microscopic body but even the shape of organic molecules? In Dennis's old

He called the electron diffraction pattern a "hologram" because it would contain the whole information, amplitude and phase, and the magnification would be the ratio of the optical to the electron wavelengths. With a very clever assistant, Mr. Williams, he decided to produce an optical hologram and then reconstruct the object optically.

laboratory in Charlottenburg, Knoll and Ruska began the development of the world's first electron microscope using the actual body of Dennis's old cathode-ray oscillograph. He always regretted that he had left the evolution of this great development, and he always retained a very deep attachment to electron microscopy.

Hitler came to power in 1933, and Siemens did not renew Gabor's contract of service, so he returned to Hungary and from there wrote to me asking if I could help him find a job in England. My old professor, Lord Rutherford, and Einstein had launched an appeal in the Albert Hall in London to help foreign scientists escaping from Hitler to gain employment in Britain, and my director of research in Manchester was prepared to help Dennis to come to work with me; however, as Dennis was working on plasma lamps, the best place to work was in our lamp factory in Rugby, and there he came in 1934. For several years he worked on different varieties of lamps, but he was full of ideas of all kinds and turned his mind to other requirements of the company. For example, he invented an extremely complicated system of three-dimensional picture projection for use in cinemas; it was based on very advanced optics and was such that the members of the audience did not need to wear any special form of colored spectacles. Then he invented a clever system of speech compression so that the transatlantic cables would be able to carry more channels so that speech could be simultaneous. But above all, he met his wife. In the musical circles he joined in Rugby was a charming girl, Marjorie Butler; by 1936 they were married, and as Dennis wrote many years later, "They lived happily ever after." Indeed, they did. They were ideally suited to each other; she was very practical, a good cook and hostess, a perfect companion to a scientist who often had his head in the clouds. She accompanied him on many of his travels and set up beautiful homes in London and later in Italy.

In my laboratory in Manchester, we made the first British electron microscope in 1934/35, and Dennis kept in touch with our work. He wrote several important papers on electron optics, and he also wrote the first book on the electron microscope, which was published during World War II. In the last chapter, he wrote that he was still hoping to see single atoms, to improve, somehow, the resolution of the electron microscope beyond the limit set by chromatic and spherical aberrations, a limit calculated by Schertzer at about 4 Angstroms. He invented a zonally corrected lens based on the introduction of space-charge; a filament threading centrally through the lens would emit electrons. We looked very carefully at this invention, but decided it was an impractical concept; he agreed and continued to think of other alternatives in his search for a solution. He had studied Zernicke's use of a coherent background of waves to show up aberrations in optical lenses and mirrors, and he had been to Cambridge to see Bragg's apparatus for structure analysis by interference of diffracted beams from millions of unit cells, so his mind was attuned to the optics of scattering and the interference of one beam with a coherent background of waves. Suddenly, as he was awaiting his turn to play tennis on Easter 1947, the solution presented itself to him in a flash; his subconscious had been working for months on the problem, and it delivered up the solution on a platter. "Why not take an electron microscope picture, one which contains the whole information, and correct it by optical means; to capture the whole information including the phase, the coherent background must be supplied by the same electron beam, which will therefore produce interference fringes; photograph these and then illuminate this photograph with light and focus it onto another photographic plate." He called the electron diffraction pattern a "hologram" because it would contain the whole information, amplitude and phase, and the magnification would be the ratio of the optical to the electron wavelengths. With a very clever assistant, Mr. Williams, he decided to produce an optical hologram and then reconstruct the object optically. By 1948, they had produced a very good reconstruction, and Dennis wrote up the full mathematical analysis, which has well stood the test of time; it contains all the necessary information on which holography is based. We worked very hard to produce electron holograms in the electron microscope with his support but were never quite successful, and after a few years we had to give up. Dennis lost all interest in the subject, and for 15 years nothing more was ever heard of electron microscopy by reconstructed wavefronts. Then the laser was born in 1960.

Dennis left industry and was appointed to Imperial College in 1948, first as a reader and later as a professor, and for 20 years he greatly enjoyed academic freedom. He had a small research school in electron physics and gave his students some formidable problems, all of them based on his inventions. Unlike the nuclear physicist, he was not a discoverer of new phenomena; he was an inventor of things and processes. Holography had contained nothing new; it was based on the optics of the nineteenth century; it merely had to be invented. He invented a flat television tube; in it three electron beams were first brought down from their guns, then deflected through 180 degrees, passing upwards, and then bent through 90 degrees so that they struck the multicolored dots of the phosphor on the front surface of the tube. It was a lovely concept, and many students in turn worked on the development, but in the end it had to be dropped for it was just too difficult a task for a small university laboratory to undertake. He then entered the field of thermonuclear fusion with some bright ideas but in competition with the huge teams at Los Alamos, Livermore, Princeton, and England, and after a time this too proved to be too difficult for students to tackle. He followed his students' work with the eye of a lynx, giving fresh instructions at every step, but the instructions bordered on the limit of practicality for students to handle, and several more exciting ventures had to be dropped. What was almost his last work concerned a brilliant concept for a thermoelectric-generator.

In 1962 he published a small book called *Inventing the Future*. He asked himself, "What future can I invent which would make for happiness whilst meeting the requirements demanded by advancing technology? For the first time in history we are faced with the possibility of a world in which only a minority need to work to keep the great majority in idle luxury." It is one of the best books ever written on the way the next generation *ought to travel,* and it was translated into several languages.

Retirement came at last in 1967, and he had Marjorie set up their lovely home near Anzio, south of Rome, where they entertained with wonderful generosity.

Once the laser had been developed, holography galloped ahead; by 1971 more than 2,000 scientific papers on the subject had been published, and attention was focused on the fundamental work of the man who had invented holography way back in 1947/48. His name must have been before the Nobel Prize Committee, but for the past four years the nuclear physicists had won the prizes, and Dennis sadly reflected that the days of the inventor were over. So the Nobel in 1971 came as a very great surprise and gave great pleasure to the host of friends who had admired his many achievements. Modestly he said, in the Nobel lecture, "I am one of the few lucky men who could see an idea of theirs grow into a sizeable chapter of physics. I am deeply aware that this has been achieved by an army of young, talented, and enthusiastic researchers, and I want to express my heartfelt thanks to them for having helped me by their work to win this greatest of scientific honors; I am surprised that I have obtained the Nobel for such a simple invention." Simple it might have been—indeed, the concept was simple—but a study of his papers of that time shows how immense was his mathematical ability and his great insight in dealing with the wave theory of optics.

He had retired to Italy when the Club of Rome was founded by a group of men of various disciplines who were all greatly concerned with the future of the world in the face of galloping technology, and he wrote two more books, *The Mature Society* and *Beyond the Age of Waste,* which dealt with these weighty matters. He was also employed by the CBS in America for many months of the years, his inventive genius flowering afresh with the application of the laser to holography. One development gave him great pleasure: by incorporating a few heavy atoms into a crystalline lattice it had been possible to see these atoms with the electron microscope by his original holographic technique; the wheel had come full circle. Alas, only just in time, for even as he was writing up this work, his last paper, he was struck down by illness. He recovered sufficiently to travel. Marjorie was marvelous in attending to him, but he never recovered speech, nor could he read or write. His bravery elicited admiration from all who met him. By 1978 he was housebound. My wife and I saw him in January 1979; he understood all we said, but he was very weak and died within the month.

Dennis called himself an "ideas man"; this he was par excellence. He quickly saw through a problem, applied what he called "a little mathematics," (but his idea of a little mathematics was far from the general concept!) and the problem was solved. He was quite one of the most stimulating men any of us has been fortunate enough to call a friend.

Emil Wolf

Recollections of Max Born

This article is essentially the text of lectures presented September 7, 1982 at the Max Born Centenary Conference held in Edinburgh, Scotland and October 21, 1982 at the Max Born Symposium held during the Annual Meeting of the Optical Society of America.

Dennis Gabor.

Max Born at his desk, ca. 1950. (Credit: AIP Niels Bohr Library).

The invitation to address this commemorative meeting has given me the rare opportunity to set aside my customary activities and try to recall a period of my life several decades ago when I had the great fortune of being able to collaborate with Max Born. As the title of my talk suggests, this will be a rather personal account, but I will do my best to present a true image of a scientist who has contributed in a decisive way to modern physics in general and to optics in particular; it will also present glimpses of a man who, under a somewhat brusque exterior, was a very humane and kind person and who in the words of Bertrand Russell was brilliant, humble, and completely without fear in public utterances.

The early part of my story is closely interwoven with another great scientist, Dennis Gabor, through whose friendship I became acquainted with Born.

I completed my graduate studies in 1948 at Bristol University. My PhD thesis supervisor was E.H. Linfoot, who at just about that

Emil Wolf is Wilson Professor of Optical Physics at the University of Rochester. Reprinted from *Optics News*, November/December 1983 with permission from the author and the Optical Society of America.

time was appointed Assistant Director of the Cambridge University Observatory. He offered me, and I accepted, a position as his assistant in Cambridge. During the next two years while I worked in Cambridge I frequently traveled to London to attend the meetings of the Optical Group of the British Physical Society. They were usually held at Imperial College and were often attended by Gabor, whose office was in the same complex of buildings. From time to time I presented short papers at these meetings. At the end of some of the meetings Gabor would invite me to his office for a chat. He would comment on the talks, make suggestions regarding my work, and speak about his own researches. Gabor liked young people, and he always offered encouragement to them. He knew Born from Germany, and he had great admiration for him.

Through Gabor I learned in 1950 that Born was thinking of preparing a new book on optics, somewhat along the lines of his earlier German book *Optik,* published in 1933, but modernized to include accounts of the more important developments that had taken place in the nearly 20 years that had gone by since then. At that time Born was the Tait Professor of Natural Philosophy at the University of Edinburgh, a post he had held since 1936, and in 1950 he was 67 years old, close to his retirement. He wanted to find some scientists who specialized in modern optics and who would be willing to collaborate with him in this project. Born approached Gabor for advice, and at first it was planned that the book would be written jointly by him, Gabor, and H. H. Hopkins. The book was to include a few contributed sections on some specialized topics, and Gabor invited me to write a section on diffraction theory of aberrations, a topic I was particularly interested in at that time. Later it turned out that Hopkins felt he could not devote adequate time to the project, and in October of 1950, Gabor, with Born's agreement, wrote to Linfoot and me asking if either of us, or both, would be willing to take Hopkins' place. After some lengthy negotiations it was agreed that Born, Gabor, and I would co-author the book.

The start of collaboration

I was, of course, delighted with this opportunity, but there was the problem of my finding the necessary time to work on this project while holding a full-time appointment with Linfoot at Cambridge. I mentioned this to Gabor, and I told him that if there were any possibility of obtaining an appointment with Born, which would allow me to spend most of my time working on the book, I would gladly leave Cambridge and go to Edinburgh.

The building on Drummond Street in Edinburgh that housed Max Born's Department of Applied Mathematics.

Gabor took up the matter with Born, who was interested. Toward the end of November 1950, Gabor wrote me that Born would be in London a few days later and that he (Gabor) was arranging for the three of us to meet the following weekend. It was agreed that I would come to Gabor's office at Imperial College on the following Saturday morning, December 2, 1950, and that we would then go to his home in South Kensington, within walking distance of Imperial College. Born was to come directly to Gabor's home from his London hotel, and the three of us and Mrs. Gabor would have lunch there.

I arrived at Gabor's office just before lunch, and I have a vivid recollection of that meeting. There was a long staircase leading to the entrance hall of the building. As we were walking down the staircase, Gabor suddenly became somewhat apprehensive. He knew that our luncheon meeting might lead to an appointment for me with Born, and he said to me, "Wolf, if you let me down, I will never forgive you. Do you know who Born's last assistant was? Heisenberg!" This statement was not accurate. Born had other assistants after Heisenberg, but the remark shows how nervous Gabor was on that particular occasion. Fortunately, all turned out well, and Gabor certainly seemed in later years well satisfied with the consequences of our luncheon with Born.

During that meeting Born asked me a few questions, mainly about my scientific interests, and before the lunch was over he invited me to become his assistant in Edinburgh, subject to the approval of Edinburgh University. It seemed to me remarkable that Born should have made up his mind so quickly, without asking for even a single letter of reference, especially since I had published only a few papers by that time and was quite unknown to the scientific community.

Later, when I got to know Born better, I realized that his quick decision was very much in line with one trait of his personality; he greatly trusted the judgment of his friends. Since Gabor recommended me, Born considered further inquiries about me to be superfluous. Unfortunately, as I also learned later, Born's implicit trust in people whom he considered to be his friends was occasionally misplaced and sometimes created problems for him.

A few days after our meeting I received a telegram from Born inviting me to a formal interview at Edinburgh University. The interview took place about two weeks later, and the next day Born wrote me saying that the committee which interviewed me recommended my appointment as his private assistant, beginning January 23, 1951. I resigned my post in Cambridge and took up the new appointment. Later I learned that committee approval was not really needed because my salary was to be paid from an industrial grant that was entirely at Born's disposal. However, on this occasion Born was careful, because some time earlier he had had on his staff Klaus Fuchs, who turned out to be a spy for the Russians, and Born got rather bad publicity from that.

Now, the name Fuchs means fox in German, and before inviting me to Edinburgh, Born apparently wrote to Sir Edward Appleton, the Principal of Edinburgh University at that time, saying that he felt the decision about this particular appointment should not be made by him alone; since he would like to appoint a Wolf after a Fox!

Arrival at Edinburgh

I arrived in Edinburgh toward the end of January 1951, eager to start on our project. Born's Department of Applied Mathematics was located in the basement of an old building on Drummond Street. I was surprised by the small size of the department. Physically it consisted of Born's office; an adjacent large room for all of his scientific collaborators, about five at that time; a small office for Mrs. Chester, his secretary; two rooms for the two permanent members of his academic staff, Robert Schlapp, a senior lecturer, and Andrew Nisbet, a lecturer; and one lecture room. The rest of the building was occupied by experimental physicists under the direction of Professor Norman Feather. In earlier days the building housed a hospital, in which Lord Lister, a famous surgeon known particularly for his work on antiseptics, also worked.

In spite of his advanced age Born was very active and, as throughout all his adult life, a prolific writer. He had a definite work routine. After coming to his office he would dictate to his secretary answers to the letters that arrived in large numbers almost daily. Afterward he would go to the adjacent room where all his collaborators were seated around a large U-shaped table. He would start at one end of it, stop opposite each person in turn, and ask the same question: "What have you done since yesterday?" After listening to the answer he would discuss the particular research activity and make suggestions. Not everyone, however, was happy with this procedure. I remember a physicist in this group who became visibly nervous each day as Born approached to ask his usual question, and one day he told me that he found the strain too much and that he would leave as soon as he could find another position. He indeed did so a few months later. At first I too was not entirely comfortable with Born's question, since obviously when one is doing research and writing there are sometimes periods of low productivity. One day when Born stood opposite me at the U-shaped table and asked, "Wolf, what have you done since yesterday?" I said simply, "Nothing!" Born seemed a bit startled, but he did not complain and just moved on to the next person, asking the same kind of question again.

Born was always direct in expressing his views and feelings, but he did not mind if others did the same, as this small incident indicates. There will be more examples of this later.

Work at Edinburgh

We started working on the optics book as soon as I came to Edinburgh. It was understood right from the beginning that Born's main contribution would consist of making material available from his German *Optik,* but he was to take part in the planning of the new book, make suggestions, and provide general advice. Most of the actual writing was to be done by Gabor and me and a few contributors. However, like Hopkins earlier on, Gabor soon found it difficult to devote the necessary time to the project, and it was mutually agreed that he would not be a co-author after all, but would just contribute a section on electron optics. So in the end it became my task to do most of the actual writing. Fortunately I was rather young then, and so I had the energy needed for what turned out to be a very large project. I was in fact 40 years younger than Born. This large age gap is undoubtedly responsible for a question I am sometimes asked—whether I am a son of the Emil Wolf who co-authored *Principles of Optics* with Max Born!

Although I did most of the writing, Born read the manuscript and made suggestions for improvements. We signed a contract with the publishers about a year after I came to Edinburgh, and we hoped to complete the manuscript by the time Born was to retire, one-and-a-half years later. However, we were much too optimistic. The writing of the book took about eight years altogether.

Throughout his life Born was a quick, prolific writer, publishing more than 300 scientific papers, about 31 books (not counting different editions and translations), apart from numerous articles on nonscientific topics.[1] In spite of my relative youth I could not compete with the speed with which Born wrote, even at his advanced age, and it soon became clear to me that he was not too pleased with my slow progress.

One day when I started writing an Appendix on Calculus of Variations, Born said that the best treatment of that subject he knew of was in his notes of lectures given by the great mathematician David Hilbert in Göttingen in the early years of this century. Born suggested that he dictate the Appendix to me, following in the main

Hilbert's presentation, and that we acknowledge this in the preface to our book. So we started in this way. After each dictating session I was to rewrite the notes and give them to Born the next day for his comments. But we did not get very far this way. After about two dictating sessions Born said he could prepare the whole Appendix himself much faster without my help, which he then did. It is essentially in this version, written by Born, that the Appendix on Calculus of Variations appears in our book.

Born's revered teacher

Incidentally, David Hilbert, whose presentation Born closely followed, was one of Born's great heroes. To physicists Hilbert is mainly known in connection with the concept of the Hilbert space and as co-author of the classic text *Methods of Mathematical Physics,* referred to generally as "Courant-Hilbert." But Hilbert contributed in a fundamental way to many branches of mathematics and was generally considered to have been the greatest mathematician of his time. Born became acquainted with Hilbert soon after coming to Göttingen in 1905, later becoming Hilbert's private assistant. In one of his later writings Born refers to Hilbert as his "revered teacher and friend," and in a biography of Hilbert by Constance Reid,[2] published in 1970, Born is quoted as saying that his job with Hilbert was to him "precious beyond description because it enabled [him] to see and talk to him every day."

Born had an encyclopedic knowledge of physics and whatever problem one brought to him, he was able to offer a useful insight or suggest a pertinent reference. He also knew personally all the leading physicists of his time and would often recall interesting stories about them.

Optics in those days—remember we are talking about optics in pre-laser days—was not a subject that most physicists would consider exciting; in fact, relatively little advanced optics was taught at universities in those days. The fashion then was nuclear physics, particle physics, high energy physics, and solid state physics. Born was quite different in this respect from most of his colleagues. To him all physics was important, and rather than distinguish between "fashionable" and "unfashionable" physics he would only distinguish between good and bad physics research.

Born was equally broad-minded about the techniques used by physicists in their research. For example,

David Hilbert, 1912.

when we were writing a section on certain mathematical methods needed to evaluate the performance of optical systems, we found that although the results given in a basic paper on this subject were correct, the derivation contained serious flaws. I was rather indignant about this, but Born just said something like, "In pioneering work everything is allowed, as long as one gets the right answer. Real justification can come later."

One of the earliest occasions when many physics students encounter Born's name comes when they start studying quantum theory of scattering. Here they soon learn about the Born approximation. This term also occurs in many of the papers on potential scattering that have been published in the more than half a century that has gone by since Born wrote a basic paper on this subject. Yet Born was rather irritated when the Born approximation was mentioned. He once said to me, "I developed in that paper the whole perturbation expansion for the scattered field, valid to all orders, yet I am only given credit for the first term in that series!"

Resistance to new discoveries

It was not always easy for Born's collaborators to convince him quickly of new discoveries. Let me illustrate this by an example from my own experience. In the early 1950s I became very interested in problems of partial coherence. One day I found a result in this area of optics that seemed to me remarkable. I phoned Born from my home one morning, told him I had an exciting new result, and asked him for an appointment to discuss it. We arranged to have lunch together that day.

Left: Max Born as Privatdozent *in Göttingen. (Reproduced from* Hilbert *by Constance Reid, ref. 2).*

Right: Max Born in the 1920s.

When I came to his office just before lunch, Born wanted to know straight away what the excitement was all about. I told him I had found that not only an optical field, but also its coherence properties, characterized by an appropriate correlation function (now known as the mutual coherence function), are propagated in the form of waves. Born looked at me rather skeptically, put his arm on my shoulder and said, "Wolf, you have always been such a sensible fellow, but now you have become completely crazy!" Actually after a few days he accepted my result, and I suspect he then no longer doubted my sanity.

This incident illustrates a fact well known to Born's collaborators—that Born had a certain resistance to accept new results obtained by others. Nonetheless, he continued thinking about them, and if they were correct he would eventually apologize for doubting them in the first place.

This trait of Born's personality is very well described by the Polish physicist Leopold Infeld, who collaborated with Born in Cambridge in the 1930s. I will quote shortly some very perceptive observations Infeld made about Born in his biography[3]; but before doing so I would like to mention a small incident relating to this book.

One day I browsed through a bookstore in Edinburgh and found a used copy of Infeld's book. I was astonished to note that the book had Born's signature on its first page. I purchased it and asked Born the next day whether he knew the book. He said, "Yes, I had a copy of it and there is a funny description of me in it; but I lent it to someone and it was never returned. I cannot remember whom I lent it to." The book I had purchased was obviously Born's missing copy, so I gave it to him, much to his delight.

In the book Infeld describes some of his experiences in Cambridge. He started working with Dirac but found him rather uncommunicative. Later Infeld attended some of Born's lectures. During one of them Born gave an account of some results that he had recently obtained. Infeld could not understand one of Born's arguments. He borrowed his notes so that he could study the argument more closely later. Let me now quote from Infeld's biography [ref. 3, p. 176 *et seq.*]:

On the evening of the day I received the paper the point suddenly became clear to me. I knew that the mass of the electron was wrongly evaluated in Born's paper and I knew how to find the right value. My whole argument seemed simple and convincing to me. I could hardly wait to tell it to Born, sure that he would see my point immediately. The next day I went to him after his lecture and said:

"I read your paper; the mass of the electron is wrong." Born's face looked even more tense than usual. He said: "This is very interesting. Show me why." Two of his audience were still present in the lecture room. I took a piece of chalk and wrote a relativistic formula for the mass density. Born interrupted me angrily:

"This problem has nothing to do with relativity theory. I don't like such a formal approach. I find nothing wrong with the way I introduced the mass." Then he turned toward the two students who were listening to our stormy discussion. "What do you think of my derivation?" They nodded their heads in full approval. I put down the piece of chalk and did not even try to defend my point. Born felt a little uneasy. Leaving the lecture room, he said, "I shall think it over."

Infeld then goes on to say:

I was annoyed at Born's behavior as well as at my own and was, for one afternoon, disgusted with Cambridge. I thought: "Here I met two great physicists. One of them does not talk. I could as easily read his papers in Poland as here. The other talks, but he is rude." . . . The next day I went again to Born's lecture. He stood at the door before the lecture room. When I passed him he said to me: "I am waiting for you. You were quite right. We will talk it over after the lecture. You must not mind my being rude. Everyone who has worked with me knows it. I have a resistance against accepting something from outside. I get angry and swear but always accept it after a time if it is right."

Our collaboration had begun with a quarrel, but a day later complete peace and understanding had been restored.

Left: Mrs. Hedwig Born, 1961.
Above: The house of Max and Hedwig Born in Edinburgh, at 84 Grange Loan.

A little further on in his biography, Infeld speaks about Born again, and this is what he says:

I marveled at the way in which he managed his heavy correspondence, answering letters with incredible dispatch, at the same time looking through scientific papers. His tremendous collection of reprints was well ordered; even the reprints from cranks and lunatics were kept, under the heading "Idiots." Born functioned like an entire institution, combining vivid imagination with splendid organization. He worked quickly and in a restless mood. As in the case of nearly all scientists, not only the result was important but the fact that he had achieved it.

Infeld later continues:

There was something childish and attractive in Born's eagerness to go ahead quickly, in his restlessness and his moods, which changed suddenly from high enthusiasm to deep depression. Sometimes when I would come with a new idea he would say rudely, "I think it is rubbish," but he never minded if I applied the same phrase to some of his ideas. But the great, the celebrated Born was as happy and as pleased as a young student at words of praise and encouragement. In his enthusiastic attitude, in the vividness of his mind, the impulsiveness with which he grasped and rejected ideas, lay his great charm.

I regard these remarks of Infeld as a true and very perspective description of Born's mode of work and of Born's personality.

Kind and compassionate

In spite of Born's occasional irritation and impatience, he was a person who cared deeply for the well-being of his fellow scientists and collaborators. His wife, Hedwig Born, was likewise a person with deep compassion for others. She too was a remarkable and gifted person. Mrs. Born published a number of books, especially poetry, and around 1938 became a Quaker, remaining active in the Quaker movement for the rest of her life.

I would like to give just one example from my own experience, which illustrates Born's concern for others. A few months after I began working with Born, I was getting married. In those days it was difficult to rent an apartment in Edinburgh. One day during the time when we were searching for a home I received a letter from Mrs. Born, who was then with Professor Born on a visit to Germany. She said that they had heard about our problem and were very concerned that we might have to postpone getting married if we did not find somewhere to live. She then offered to help us, suggesting that we share with them their small house in Edinburgh.

Albert Einstein in the 1920s. (Credit: AIP Niels Bohr Library).

In the end we found an apartment elsewhere; but this small episode is an indication of the warmth of their personalities and of their willingness to make a personal sacrifice to help, when help was needed.

I mentioned earlier, that one of Born's great heroes was the mathematician David Hilbert. But there was another, even greater hero in Born's life: Albert Einstein, with whom he and also Mrs. Born maintained close personal friendships for almost half a century. Unfortunately, after Einstein left Europe for America in 1932 they did not see each other again, but they carried on extensive correspondence until Einstein's death in 1955. The letters they exchanged were published in 1971, together with Born's commentary, and the volume[4] is a precious contribution to the history of physics and of the times in which they lived.

There is an episode I would like to relate briefly in connection with Born's friendship with Einstein. In the early 1950s, when Sir Edmund Whittaker was preparing the second volume of his classic work *A History of the Theories of Aether and Electricity*, he sent Born the manuscript of a section dealing with the special theory of relativity. Whittaker's treatment placed a heavy emphasis on the work of Poincaré and Lorentz and dismissed Einstein's contribution as being of rather minor significance. Born, who himself wrote a book on the theory of relativity, was most unhappy with Whittaker's manuscript and sent him a long report in which he analyzed in detail the various contributions, indicating why he considered Einstein's contribution to be much more fundamental.

However, Born did not succeed in changing Whittaker's opinion.[5] In September of 1953, around the time Whittaker's book was published, Born wrote to Einstein about this. Let me quote from Born's letter:[6] "Many people may now think (even if you do not) that I played a rather ugly role in this business. After all it is common knowledge that you and I do not see eye to eye over the question of determinism."

Einstein was not concerned. This is what he said in his reply to Born:[7] "Don't lose any sleep over your friend's book . . . . If he manages to convince others, that is their own affair. I myself have certainly found satisfaction in my efforts. . . ." and then Einstein added, "After all, I do not need to read the thing."

Born retired that year (in 1953). The accompanying photograph shows Born with the members of his department at the time of his retirement.

Sir Edmund Whittaker. (Reproduced by the courtesy of the University of Edinburgh).

Life in retirement

Soon afterward the Borns left Edinburgh and settled in Bad Pyrmont, a spa in West Germany, not far from Göttingen, where they built a small house. When they left Edinburgh our book was far from finished. We corresponded about it, and I visited Born in his new home several times. Born was hoping that he and Mrs. Born would be able to lead a more quiet life in Bad Pyrmont, but he told me on one of my visits that this proved difficult to achieve. For example, soon after they settled in Bad Pyrmont, Born was invited to address a meeting of a West German physical society. He declined the invitation, saying that he was too old to travel. He received a reply stating that in view of this the meeting would be moved to Bad Pyrmont!

In 1954, the year after his retirement, Born was awarded the Nobel Prize. He was, of course, delighted, but I am quite sure he felt, as many others did, that this great recognition had come somewhat late. The Nobel Prize was awarded to him for contributions that he made almost 30 years earlier. However, as his son Gustav later noted in a postscript to Born's memoirs[8], it came at the right time to add weight to his main retirement occupation, which was to educate thinking people in Germany and elsewhere in the social, economic, and political consequences of science and also of the dangers of nuclear weapons and re-armament.

In 1957 I was a Visiting Scientist at the Courant Institute of New York University, still working on our book. One day I received a letter from Born asking me why the book was not yet finished. I

Members of Max Born's department at the time of his retirement (1953) from the Tait Chair of Natural Philosophy at the University of Edinburgh. Standing (left to right): E. Wolf, D. J. Hooton, A. Nisbet. Sitting: Mrs. Chester (secretary), M. Born, R. Schlapp.

Reprinted from New Scientist, January 13, 1966, Pages 74-78

Dear Emil,

I consider you as my chief Prophet!

Love

Dennis

HOLOGRAPHY, OR THE "WHOLE PICTURE"

By Professor Dennis Gabor, FRS
Department of Electrical Engineering, Imperial College, London

A dedication from Dennis Gabor

replied that practically the whole manuscript was completed, except for a chapter on partial coherence on which I was still working. Born wrote back almost at once, saying something like, "Who apart from you is interested in partial coherence? Leave that chapter out and send the rest of the manuscript to the printers." Actually I completed that chapter shortly afterward and it was included in the book.

It so happened that within about two years after the publication of our book (in 1959) the laser was invented and optical physicists and engineers then became greatly interested in questions of coherence. Our book was the first that dealt in depth with this subject, and Born was then as pleased as I was that the chapter was included.

Our book was also one of the first textbooks containing an account of holography. Gabor was very happy about it. Later, when holography became popular and useful, he sent me a reprint of one of his papers with a charming dedication (see above).

As I approach the end of my reminiscences about Max Born,

Max Born in Bad Pyrmont feeding pigeons.

Max Born with two of his grandsons, Max and Sebastian (children of his son Gustav) in Bad Pyrmont. (Credit: AIP Niels Bohr Library).

I would like to say that I hope my talk conveyed to you the warmth and the affection with which he remains in my memory—not only as a great scientist, but also as a kind and remarkable human being. My feelings about our collaboration are well described by exactly the same words that Born used when he spoke about his association with David Hilbert, quoted earlier, namely that my appointment with him was precious to me beyond description, because it enabled me to see and to talk to him every day.

Olivia

Before ending I would like to show you a few pictures taken in Bad Pyrmont during Born's retirement and also to mention one more episode. One shows Professor and Mrs. Born with one of their daughters, Irene. Some years ago I learned that Irene is the mother of a lady who has achieved fame comparable to that of Max Born himself, but in an entirely different field. I am speaking of the pop singer Olivia Newton-John. Shortly after I learned that Olivia Newton-John was Max Born's granddaughter, I was on a sabbatical leave at the University of Toronto. Olivia was scheduled to give a concert in Toronto while I was there. I wrote to her, told her I had collaborated with her grandfather in the writing of a book, and asked her whether we could meet. I received a charming reply in which she invited me to meet her after the concert. We met then and talked mainly about her grandparents. Before I left Olivia gave me two autographed photos of herself. Let me add that to some of my students I am known not so much as the co-author of *Principles of Optics* but rather as the person who knows Olivia Newton-John

At right, Max Born in front of his library at his home in Bad Pyrmont.

Far right: Hedwig Born and Max Born, with their daughter Irene Newton-John in Bad Pyrmont, 1957. (Credit: AIP Niels Bohr Library).

Left: Olivia Newton-John, granddaughter of Max Born.

Right: Max Born. (Credit: Lotte Meitner-Graf).

and who has a picture of her hanging in his office signed "To Emil, Love, Olivia."

I cannot bring you the voice of Max Born, but I will end my presentation with one of the songs that made Olivia famous. (The lectures on which this article is based concluded with an excerpt from the song "If You Love Me Let Me Know.")

1. Bibliography of Born's scientific publications is given in "Max Born," by N. Kemmer and R. Schlapp in *Biographical Memoirs of the Royal Society*, 17, London: the Royal Society, 1971, pp. 17-52. Born's autobiography cited as ref. 8 below was published posthumously, first in German in 1975 and is, therefore, not included in that bibliography.

2. C. Reid: *Hilbert* (Springer-Verlag, New York, Heidelberg, Berlin, 1970), p. 103.

3. L. Infeld: *Quest*, The Evolution of a Scientist (Doubleday, Doran and Co., New York, 1941)

4. *The Born-Einstein Letters*, with commentaries by Max Born (Walker and Company, New York, 1971).

5. Born's opinion on this question rather than Whittaker's is generally accepted. See, for example, D. Martin's biographical note about E. T. Whittaker in *Dictionary of Scientific Biography*, C. C. Gillespie, editor-in-chief (Charles Scribner's Sons, New York, 1976), Vol. XIV, p. 317; or A. Pais: *Subtle Is the Lord, The Science and the Life of Albert Einstein* (Clarendon Press, Oxford, and Oxford University Press, New York, 1982), p. 168.

6. Ref. 4, p. 197.

7. Ref. 4, p. 199.

8. Max Born: *My Life* (Taylor and Francis, London, and Charles Scribner's Sons, New York, 1978), p. 296.

Acknowledgments: In preparing this article for publication I received assistance with obtaining some of the photographs, determining the approximate dates when they were taken and with checking some of the references. I am particularly obliged to G.V.R. Born (London University), R.M. Sillitto and S.D. Fletcher (University of Edinburgh), L.H. Caren (University of Rochester), and D. Dublin (American Institute of Physics) for their help.

Colonel Murray Green, USAF-Ret.

George Goddard: The Father of U.S. Aerial Reconnaissance

Few Americans have contributed more to America's national security in times of war and international crisis than Brigadier General George William Goddard, who died in October 1987, only 20 months short of his 100th birthday on June 15, 1989. He is especially mourned by "Goddard's Air Force," an informal group of some 300 admirers who worked with or knew General Goddard during his 45-year service career.

George Goddard stands beside a DH-4 at Bolling Field where he was chief photo officer in 1922. (Photo courtesy Diane Bergh.)

One of the authentic geniuses of our time, Goddard patented many inventions in aerial photography. Five of his major innovations could have made him a millionaire many times over, but he chose to endow them to the U.S. government.

Like many original thinkers, Goddard didn't have the time and bureaucratic skills needed to cope with the infighting inevitable in a large military organization at war. For a time, General H. H. "Hap" Arnold, commanding the Army Air Forces, lost confidence in him and relegated him to a demeaning assignment. Goddard's services might have been lost for the duration of World War II had it not been for a Hollywood movie mogul and a senior U.S. Naval officer who restored him to a responsible role appropriate to his special skills.

In 1904, Goddard immigrated to America as a teenager from Tunbridge Wells, Kent, England. When the United States went to war in April 1917, he enlisted in the Army Aviation Section, Signal Corps, aspiring to be a flier. After basic training, he logged spruce trees in the state of Washington. Spruce was then used to make airplane fuselages.

It took a year for him to work his way into flying school. Then, having qualified for his wings, he confessed to a strong interest in photography. This resulted in a further detour to a professional school at Cornell University. Goddard never did get overseas to fulfill his original goal of dueling with the Red Baron and other aerial devils.

One day in February 1919, Goddard was puttering round inside an airplane cockpit. He found himself looking up at Brigadier

Reprinted from *The Retired Officer Magazine*, March 1990.

General Billy Mitchell, who inquired what he was doing. "I'm just rigging up this camera mount on tennis balls, sir," Goddard replied. "I thought they might make a good shock absorber."

That and other improvisations brought then Lieutenant Goddard to the attention of Colonel Edward Steichen, the world-famous photographer who was in charge of the Army's wartime aerial reconnaissance program. Before Colonel Steichen left the service, he helped Goddard, who became chief of aerial photography at McCook Field (later Wright-Patterson AFB), Ohio.

In 1921, aviation pioneer General Mitchell, having a need for clear, precise aerial photography, dragooned Goddard to fly with him off the Virginia capes to capture on film the epochal photos of the destruction of the "unsinkable" battleship Ostfriesland and two other captured German warships. Still later, Goddard took the first night aerial pictures while stationed at the Air Service Photographic School, Chanute Field, Illinois.

In 1934, Goddard achieved still another conceptual first. "Hap" Arnold, commanding 10 spanking-new Martin B-10 bombers, had been ordered by the War Department general staff to fly a simulated combat mission to Alaska and back to Washington, D.C., to demonstrate to a skeptical Congress and nation that the air component of the Army was not as ineffective as a disastrous airmail mission some months before had seemed to suggest.

As official photographer, then Captain Goddard worked around the clock to set up a field laboratory. Arnold bought Goddard's idea to fly six planes abreast northward from Anchorage to Fairbanks, taking advantage of the sun's angle to produce a continuous strip photo 50 miles across over a 280-mile span. In all, 34,000 square miles of bleak Alaskan landscape were photographed, and for the first time, cameras unlocked age-old geographic and geologic secrets that promised untold potential. Many of the spectacular photos later appeared in National Geographic Magazine.

Upon successful completion of the mission, Arnold was rewarded with an invitation to visit President Franklin D. Roosevelt in the White House. But for Goddard, the job was only half done. He gathered his negatives and prints, and for weeks monopolized the entire Chanute Field base gymnasium floor. Laboriously, he fitted together thousands of pieces of a giant mosaic. The resulting maps, the first of their kind, were delivered to a grateful U.S. Geological Survey.

Those who undertook historic missions continued to seek out George Goddard's unique talents. Not quite four years later, his nosy camera recorded another revolution in military strategy. Major

**When President
Roosevelt made a
politically inspired
tour of Wright Field,
Ohio, to advance his
re-election
campaign, Goddard
set up a
photographic
display that included
a spectacular color
shot of the Roosevelt
family estate at
Hyde Park, New
York. Roosevelt
dallied for a half-
hour looking over
the impromptu
display.**

*Goddard discusses the merits and progress of aerial photography with
President Franklin Roosevelt at Wright Field, Ohio. (Courtesy Diane Bergh.)*

General Frank Andrews, then commanding general of Air Force
General Headquarters at Langley Field, Virginia, set a mission for
three B-17 Flying Fortresses, the newest bomber in the inventory.
On May 2, 1938, three "Forts" flew east from Mitchell Field, New
York, on an Atlantic Ocean search mission. Unerringly directed by
the lead navigator, 1st. Lieutenant Curtis LeMay, the bombers
emerged from a heavy cloud bank and intercepted the Italian liner
Rex 725 miles out of New York. Goddard's plane hung back to
allow him to "shoot" the other two bombers flying just above
smokestack level, startling the passengers on deck. After a circuit
or two and an invitation radioed from the liner for the air crew to
stop off for lunch, the B-17s headed back. The next morning, the
front page of the staid New York *Herald Tribune* displayed a
spectacular photograph of the intercept that was reproduced in
1,800 newspapers and magazines nationwide.

This remarkable feat, several years before the advent of radar,
conveyed to military planners a radical new strategic idea about the
potential role of airplanes in national defense, no less significant
than Mitchell's caper off the Virginia coast 17 years before.

The next morning, General Andrews' telephone rang off the
hook. Among his callers was an angry General Malin Craig, Army
Chief of Staff, who had never authorized the mission. The upshot
of the Rex interception was a verbal 100-mile limit placed on all
military aircraft flying out to sea, ostensibly for safety reasons. This
restraint was not lifted until a few months before Pearl Harbor.

Fast reputation

Goddard's reputation spread too fast for some of his supervisors. In 1940, when President Roosevelt made a politically inspired tour of Wright Field, Ohio, to advance his re-election campaign, Goddard set up a photographic display that included a spectacular color shot of the Roosevelt family estate at Hyde Park, New York. Roosevelt dallied for a half-hour looking over the impromptu display. Uncomfortable aides gently reminded the president that his schedule was backing up, all the while glaring at Goddard for his daring self-promotion.

Goddard still rated aces-high with General Arnold though. In fact, the two collaborated that year, collecting a healthy advance on a picture story featured in National Geographic. But their relationship, which had begun with the Alaska mission, eroded under attacks by bureaucratic insiders claiming that the United States had fallen behind the British in aerial reconnaissance equipment and techniques. The accuracy of this charge is hard to assess. America's "Arsenal of Democracy" was just cranking up for all-out production in 1941 to 1942. The Royal Air Force, cited by Winston Churchill as the few to whom so much was owed by so many, had earned its wings in the skies over Britain, while the fledgling U.S. Army Air Forces had yet to experience combat.

In any event, Goddard probably finally spoiled his chances of immediate advancement by committing an unpardonable sin in the eyes of the morally straight-arrow, Hap Arnold. Goddard, a married man, had fallen in love with Elizabeth Hayes, a freelance photographer and public relations employee of the Colonial Williamsburg restoration project in Virginia. At age 54, Goddard divorced his wife and married Elizabeth. Their union turned out to be a stable one, lasting 41 years until Elizabeth's death in 1984.

But that slip from rectitude convinced Arnold that Goddard was more dilettante than dexterous. A friend, who had come to Dayton, Ohio, from Washington, D.C., alerted Goddard that the ax was about to fall. "Arnold is mad at you. Don't go anywhere near him. He's looking to send you to an island in the Pacific without any palm trees." Arnold often reserved remote assignments for staff officers who, in his view, had strayed from the straight and narrow.

Goddard's orders to a small air base near Charlotte, North Carolina, floored him. His new duty: base venereal disease control officer. His exile, however, proved to be short-lived. One admirer, film producer Hal Roach, who was making training films for the Army, tracked him down and gained for him an audience with Rear Admiral DeWitt Ramsey, chief of the Bureau of Aeronautics

(BuAer). The admiral asked for a demonstration of Goddard's recently developed strip camera, which could take effective aerial photos at high speed by using the principle of image-motion compensation, in effect, moving the film to synchronize with the speed of the image while the photo was being taken. The shutterless camera produced long, continuous and often sharp photographs of the ground by wiping synchronized ground images on the film at correct speeds. According to the citation for the Distinguished Service Medal Goddard later received, his camera made exquisitely sharp photos at 500 mph and at a 50-foot altitude. It was even capable of showing railroad ties on the ground from 40,000 feet.

Strip camera's success

"We shot some test areas at Palm Beach," Goddard explained years later in an interview, "and we got within three feet of the correct depth on objects set in as much as 25 feet of water." Admiral Ramsey was excited. The Navy immediately ordered 200 strip cameras.

Brought back to Washington, D.C., Goddard was intercepted by Major General Hoyt Vandenberg, just preparing to take over the Ninth Air Force, the principal tactical air support for the forthcoming invasion of the European continent. General Vandenberg hustled his charge into the office of Robert Lovett, Assistant Secretary of War for Air, carefully bypassing General Arnold's next door.

Lovett asked Goddard to stay with the Army. In a display of individualism, George expressed his wish to join the U.S. Navy, "where they treat you right." In the end, however, he agreed to accept Army Air Force orders taking him to Great Britain. He was assigned to work with Colonel Elliott Roosevelt, commanding a photo reconnaissance group assigned to the Eighth Air Force.

Goddard also worked for the British Royal Air Force (RAF). His strip camera was installed in a Typhoon, the RAF's fastest reconnaissance plane. On one test mission, the pilot flew low over the English Channel and photographed a suspect hotel in Holland that had been "shot" many times before with conventional shutter cameras. This time, the strip camera disclosed 30 thin telephone wires fanning out of an upper window. It was clear that at last, here was the long sought Nazi regional headquarters. But not for long. RAF bombers quickly destroyed it.

Unfortunately, there was not enough time before D-Day to produce the number of strip cameras needed for the invasion. But

Goddard's cameras spotted the Soviet ship Kasimov carrying missiles in Cuban waters in 1962. (U.S. Air Force photo.)

Goddard held a clinic for his RAF counterparts and showed them how to modify their shutter cameras to obtain a comparable effect.

The Army Air Forces, having first call on Goddard's talents, used his camera with good effect to help spot underwater obstructions set by the Germans to thwart the Normandy invasion. Years later, however, Goddard reported its best success was actually achieved in April 1945 before the Okinawa invasion. Carrier-based reconnaissance determined precisely where each landing should take place. Goddard said that he was told that if his cameras had been in full use during the Leyte invasion of the Philippines six months before, casualties there could have been reduced significantly. Landing barges at Leyte had run aground on uncharted reefs, becoming easy targets for Japanese beach fire.

Fully vindicated at the end of the war, Goddard was restored to the good graces of his superiors and placed in charge of the Aerial Photo Reconnaissance Laboratory at Wright-Patterson AFB. He was promoted to brigadier general in 1951 and recalled to active duty during the Korean fracas.

He accomplished perhaps the most spectacular success of his career in 1962, while still vigorous at the age of 73. His cameras, installed in an RF-101 Voodoo reconnaissance plane, laid bare Soviet duplicity in Cuba. Clear photographic evidence of missiles being installed there gave the lie to U.S.S.R. Premier Nikita Khrushchev's vehement denials, and the famous confrontation dissolved in a face-saving withdrawal by the Soviets. In recognition of the importance of Goddard's photographs, President John F. Kennedy had a transparency of one of the best of them mounted in the Oval Office.

After that exciting climax to his career, Goddard began to slip into a well-deserved retirement. For a time, he served as a consultant to Itek Corporation, specializing in aerial reconnaissance. At last he bought a place in Boca Raton, Florida, figuring he had stayed around long enough to gain recognition for photo-optical intelligence "as an indispensable insurance policy for peace," a phrase he used in remarks at a Pentagon symposium in 1964.

In his retirement, other deserved honors came to Goddard. In July 1976 he was inducted into the Aviation Hall of Fame, whose august membership includes the Wright Brothers, Billy Mitchell, Charles Lindbergh, and others of the most select contributors to the advancement of aviation.

Looking back over his distinguished record, it is fair to state that, for more than four decades, whenever technical wizardry in aerial photography was needed, military leaders wisely "left it to George."

Rudolf Kingslake

A Lifetime in Optics

Rudolf Kingslake's career as an educator and lens designer spans more than 60 years. He taught courses in optics and lens design at the Institute of Optics at the University of Rochester since its founding in 1929, and also headed the lens design department at Eastman Kodak Company for more than three decades. He was interviewed by Robert E. Fischer, SPIE President for 1984 and editor of OE Reports.

I'd like to begin by asking about your first exposure to the field of optics.

My father was an enthusiastic amateur photographer, and I began to wonder about how the lenses on his cameras worked. He had a nice little book published by Beck, which showed sections of lenses, and I became curious about why it took six elements to make one kind of photographic lens whereas another type of lens only had four elements. So, when I heard that Imperial College had a department of optics where lens design was taught, I decided that was the obvious place for me to go. Imperial started the optics department in 1917, at the end of the war; I went there in 1921, so the department of optics was still practically in its infancy. I was in the second, full-time undergraduate class, which lasted three years.

Then you studied directly under Professor Conrady?

Professor Conrady taught the lectures on lens design. L.C. Martin taught general optics, and B.K. Johnson did the laboratory work. In fact, we'd virtually live in the laboratory, leaving it only to attend a lecture and then returning to the lab. It was basically a lab course.

You later married Professor Conrady's daughter?

Yes. That was in 1929, before we came to the U.S.

I'm wondering about the influence on your overall career of your marriage to the daughter of an optics professor.

Hilda was in the optics department also. She was one of three students in the first regular undergraduate course; there were four in our year, and two in the following year. We all worked together and got to know each other very well. There were only one or two graduate students at that time; more came later.

Ultimately, you and your wife moved to the United States. Why did

Excerpted from *Optical Engineering Reports*, January 1984.

you leave England?

Dr. Rhees, then president of the University of Rochester, came to England searching for faculty. It's all in the History of the Institute of Optics, which my wife wrote. Kodak and Bausch & Lomb were planning to help finance an Institute of Optics at the university. Dr. Rhees reckoned that there weren't any suitable faculty in this country; I don't know why. When he came to England recruiting faculty, he got me and also a man named Taylor from Cambridge. We were the first two faculty members at the Institute. A year later, O'Brien, who was an American, joined us, and he was followed by Gustave Fassin from Belgium. It was quite a mixed blend of faculty.

You started teaching at the University of Rochester in 1929?

We had only optometry students at that time, about eight of them a year. After a couple of years, optics students began to appear. It was the time of the Great Depression, and why they kept the Institute running during the depression, I have no idea. Rhees must have had plenty of faith in it. That's all in the History of the Institute.

After teaching at the university for more than 50 years, how would you compare students in earlier years to today's students?

Our early optics students were very good, and some have become quite famous. Taylor and I taught only classical optics, geometrical and physical, while O'Brien taught physiological optics. There wasn't any such thing as "modern optics," which started about 10 or 15 years ago, mostly as a result of the development of the laser. This discovery led to the study of quantum optics and coherence, with holograms, image processing, solid state detectors, and many similar fields developing as a result.

Do you think that students today derive a better understanding of the technology than they did in an earlier day, or the converse—do you see any difference?

Oh, no. Optics, to my thinking, splits into two classes—what you might call classical optics and modern optics. Classical optics students get bachelor's degrees and go into industry. They become lens designers, system designers, and instrument designers. Then there are the modern optics people, who usually come in from other universities with bachelor's degrees. They study for the PhD and then they go on to get jobs in research labs. The two are really quite separate. Master's and bachelor's students are instrument types, and PhDs are research types. We get both kinds of students at the University of Rochester.

> **Classical optics students get bachelor's degrees and go into industry. They become lens designers, system designers, and instrument designers. Then there are the modern optics people, who usually come in from other universities with bachelor's degrees. They study for the PhD and then they go on to get jobs in research labs. The two are really quite separate.**

In terms of your efforts at Kodak relative to their optical systems and camera lenses, is there any one lens that was either the most difficult or challenging to design?

No single lens stands out as most difficult or challenging. Of course, the war came along soon after I started there. Then we had to make everything. We made military instruments, range finders, telescopes, zoom telescopes for tanks, things that weren't photographic. We also made aerial camera lenses, a 36-inch, a 24-inch, a 12-inch, and the 7-inch f/2.5, which is quite famous. We had about 20 people working on lens designs at the time. New requests were always coming in from management or from the Army or from somewhere else. General Goddard [see page 30] would amble in and say, "Could you people make me a something or other?" and we'd answer, "We'll see what we can do." Fairchild took all our photographic output and put our lenses on their cameras, most of which went to the Air Force.

That sounds like a rather challenging time of your career.

It was a very challenging period because we had to do everything in a hurry. Things were wanted immediately.

Do you think the technology took a major step forward during that period?

I don't think so. What we were doing mainly involved hard work. We'd work Saturdays and nights punching buttons on desk calculators. We had no computers, nothing to help us. It was plain hard work.

What is the most significant technology advancement over the years—it's a very broad question, I know—that you have seen?

I think computers have been the major advance, but manufacturing methods have vastly improved also. At Kodak, Dr. McLeod and his group developed high-speed polishing machines, which are used extensively, and Art Simmons built a number of automatic centering and edging machines. During the war he made several large pantographs for reticle engraving, and he also developed a new type of nodal slide bench for lens testing.

Other major areas of optics technology involve lasers and fiber optics. These have been mushrooming substantially in the past years. What are your thoughts about these developments?

I don't know that much about lasers. They have been enormously successful as sources of bright coherent monochromatic radiation in parallel beams, which everybody needed and now we have. But the fiber business is a strange thing. The whole thing depended on making fibers that would transmit without absorption losses so that you could make them miles long instead of just a few feet. Once that hurdle was overcome, somebody studied the length of time it took to travel down the fiber, and they found that different paths took different times. Then they invented the parabolic cross section of refractive index and eliminated that problem. Now there seems to be no end to what fibers can do.

Do you see a time when most communication will be by optical fiber?

Yes. As opposed to wires, fibers have many advantages. You can send a carrier wave down a fiber which has the frequency of light, 10^{14} Hz, instead of carriers that go on wires, which means you can carry a dozen television signals over one carrier wave on one fiber. As for telephone messages, you could send hundreds or thousands of telephone messages on one fiber, which you couldn't possibly do over wires.

I'd like to turn our discussion back to the university and to education in optics. What about some of your laboratory projects?

One of our projects was to run a course in photography for the benefit of the students. The first thing we had to do, of course, was make a photographic plate. I found a recipe in a book, and we melted the gelatin and put in the sodium chloride and the silver nitrate. We stirred it up and steamed it for a certain time; then we chilled it and squeezed the gelatin through a cloth to make little worms and washed them to get out the soluble salts, remelted it, poured it onto a glass plate, and set it on an ice table to chill. Then we put it in a camera and went out and made a photograph. I still have the plate I made. It worked out very well. I remember Jack Tupper, who worked in the research lab at Kodak Park for years, saying that the only emulsion he ever made was that one. The worst emulsion in the class was made by a student who shall remain nameless. His had thumbprints on it, it had areas of glass where there was no emulsion, and the emulsion varied in thickness, which meant varying density. It was terrible. Most of the students did well, however, and it was quite an interesting project.

Wasn't there a requirement that students in the Institute of Optics, up until even the 60s, take an optical fabrication-related class?

Yes, and there still is something of the sort going on. It's very scrappy because there are so many students. Laboratory work is almost gone in the Institute simply because we have so many students—what can we do with 60? It's fantastic. The undergraduate geometrical optics class has to be done in two sections now

New requests were always coming in from management or from the Army or from somewhere else. General Goddard would amble in and say, "Could you people make me a something or other?" and we'd answer, "We'll see what we can do."

because it's so huge. There's not even a room big enough for them.

Do you think that people working in optical design who are coming out of today's educational process have enough understanding of how things are produced?

The only way to really understand something is to do it yourself. Imperial College had a month-long summer school class on making lenses. Actually, it was done at Northhampton Polytechnic. There we actually ground and polished lenses; it was most instructive. We made a telescope doublet and an eyepiece. My telescope doublet was very bad; the eyepiece was very good. I still have the eyepiece, which we use as a pocket magnifier at home. Although the telescope objective was terrible, we learned how it could be done, how it should be done, and how we weren't doing it. Actually, if we had test plates and had taken a little longer, we could have made quite a good lens.

What about the educational system in the United States? Let me ask specifically about higher education and the difficulty that universities are now having in attracting capable faculty.

That is a seemingly insoluble problem. The Institute of Optics is losing its faculty like mad. We lost four professors in a year and a half, leaving barely enough to handle the increasing number of students. We just cannot keep people.

What's the solution? This is certainly going to impact the quality of technical education in the United States.

One solution would be for companies like Kodak to lend the universities a man for a year. But they aren't willing to do that because they need good men. An alternative would be for the companies to finance faculty members because the university is only able to pay them about half the industrial salary. If someone could make up the difference, we might get faculty.

Isn't this something that really has to be done in order for the country to retain its technical leadership?

Yes. There are articles in the literature all the time bemoaning the fact that we can't get faculty members. The competition is too great. If the U. of R. could afford forty, fifty, or sixty thousand dollars a year to get a professor, they could compete with industry. As it is, they're paying about half that and cannot compete. The only other hope might be retirees like me who, after they leave industry, could return to the university to teach. But most retirees don't seem to want to do this. By the time they get to retirement age, they want to go and play golf in Florida.

What messages or guidance would you have for the optical

designer or optical engineer of the future?

Obviously, he's got to know everything he can, but there is much more than that. I think he will need a natural curiosity to work out the inwardness of what he's doing. Rather than simply doing a task, he's got to know why he's doing it. Yesterday, in a tutorial on high-speed cameras, I tried hard to find out the difference between a drum camera and a camera with stationary film. I talked to one man and asked, "Yes, but why does it ...?" He could not tell me. Then I talked to another man. "It's simple," he said. "You do this and then that." It all fell into place immediately. The second man understood it; the first man did not. So, I think the message is to understand what's going on all the time, even if it takes you weeks to find out.

Harold Edgerton

Strobe Photography:
A Brief History

Harold Edgerton, inventor of the electronic strobe, was a professor of electrical engineering at MIT for many years. He died on January 4, 1990. Edgerton is well known for pioneering the electronic flash lamp to stop motion of high-speed objects. He was the author of several books on high-speed photography. In this article, Edgerton discusses some of the exciting developments of the recent past in strobe photography, and relates some of the history that brought about this remarkable revolution in the photographic world.

The first known photograph taken by a flash of light from an electrical discharge (spark) was accomplished about 1850 by Henry Fox-Talbot in England shortly after he invented the negative-positive process that is used so widely today. However, strobe lighting has only recently come into its own. Just a few years ago there were few in use, and these were large, bulky, and inconvenient. Who would have predicted, in 1940, that today most miniature cameras would come equipped with a built-in strobe light, including battery, circuit, and automatic exposure capability?

The owner of a new, small strobe lamp soon will become aware of the factors that were compromised by the designers. Characteristics such as light output, flash duration, charging time, and available flashes from a battery often have been selected to achieve smallness of size and portability. I congratulate those who designed the flash units that are in such wide use today.

Several years ago, an author asked me for photos to illustrate a book about strobe lights and their uses. I gave him many examples, such as the multiflash of a golf swing and bullets in flight. To get a perfect golf swing photo took years of effort. Such a picture required two strobe lights covering a field of about 10 feet at a frequency of 120 flashes per second. The light output was designed to give a good image of a white-painted, rapidly moving golf club contrasted against a black velvet background. Another collection of photos of bullets, birds, etc., was furnished. The exposure time was less than a microsecond for the bullet and about 100 μs for the birds.

These and other photos appeared in his article without any

Excerpted from *Optical Engineering Reports*, September 1984.

Who would have predicted, in 1940, that today most miniature cameras would come equipped with a built-in strobe light, including battery, circuit, and automatic exposure capability?

explanation. I wrote to the author and suggested that he did not make it clear that the strobe units used as illustrations in his book were entirely incapable of producing the photographs that he used. I suggested further that there were no secrets about my special flash equipment and wondered why he did not include the full data for the benefit of the readers. I received a nice letter back, telling me of the pressure from his publishers to meet a publication date, and adding that he just did not have a chance to change or finish the job properly.

In 1970, I published a technical book (*Electronic Flash, Strobe,* McGraw-Hill, N.Y.) about strobe lamps, especially the special ones mentioned above, and their uses. As the years passed, the price of the book increased and sales decreased. Eventually, the sales dropped below a minimum number, and the book was discontinued. I had just written to the publisher proposing that the book be revised. Their reply stated that the book was to be taken off their list; in other words, it would be out of print.

With my urging, MIT Press picked up the pieces and published a paperback edition in 1979 using the same title. The new edition contains a new lengthy preface, many new references, updated comments, and is still being published.

There have been some unfavorable comments about this book. Electrical engineering types take a look at the book and conclude that it is all photography. Meanwhile, some photographers comment that the book is all electrical engineering in nature. It seems that it is an almost impossible task to suit everyone. Regardless, this reference, or textbook, is full of biographical material and useful information about the theory, design, and applications of electronic flash lighting equipment. I know that it has helped some of its readers and will be a source book for years to come.

At frequent intervals I get requests about the design of special strobes for specialized uses, such as bullet and bird photography. I refer such inquiries to the previously mentioned book, in which most of the answers are explained in simple terms. However, just because a circuit is drawn out completely in the book, there is no guarantee that the components are available in the marketplace. I did my best to give an accurate description of the devices, but it is up to the user to find the components.

The rules for a short-duration flash are relatively simple: (1) the watt-second input should be as small as possible; (2) the flash lamp should have a short arc of large diameter; and (3) the capacitors should have a small, internal inductance and resistance, and the connecting wires to the flash lamp should be as short as possible. There is some information about this in the above-mentioned

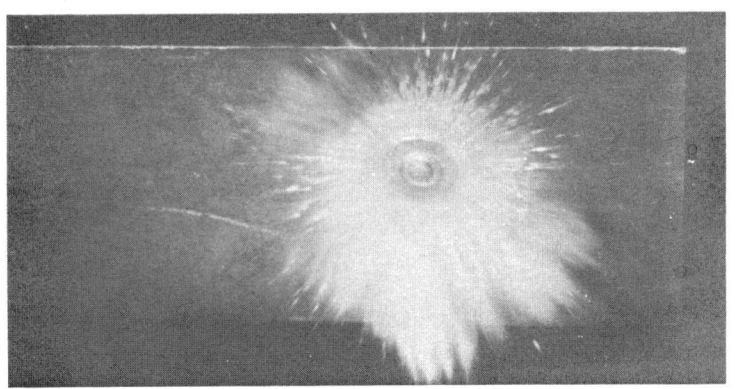

Bullet Splash (1937).
© Harold Edgerton.
Courtesy of Palm Press,
Inc.

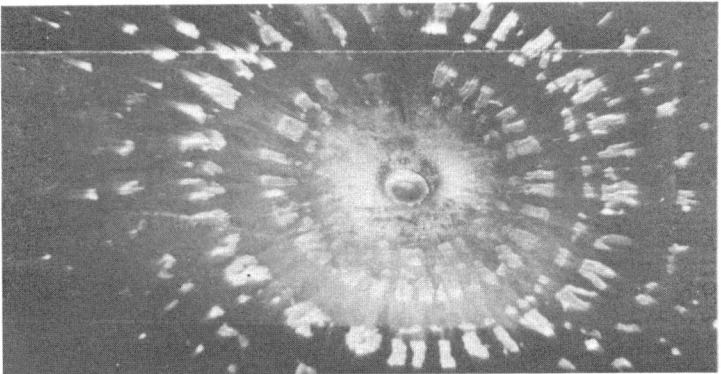

book. Further technical information is available from the manufacturers of flash lamps.

It seems to me that those who manufacture and sell electronic flash lighting equipment should make an effort to specify the beam candela power output and the flash duration. Other data, such as the number of flashes per battery charge, beam diameter, weight, charging time, etc., are also useful knowledge.

A recently published description of flash equipment, with a 25 μs flash duration, 450 bcps* output, appeared in *Photomethods* magazine (pp. 37, 40-44, Feb. 1982). It is similar to the description on p. 150 of *Electronic Flash, Strobe*. Both of the above flash equipment were designed especially for bird photography. They use quartz lamps of small size discharged from low inductance capacitors. Paper capacitors are preferred since they have a small internal resistance.

My first flash unit used a mercury-arc rectifier, with an internal grid to control the flash instant. There was a separate circuit to keep the mercury-arc spot active. The flash duration was about 10 μs, and the photographs of a synchronous machine were without blur. This strobe was mounted on a wooden framework that was nailed to the floor. It held the mercury-arc close since the light output was rather small.

As I recall, Dr. C. Stark Draper came to see the equipment. He wanted us to help him with several problems, such as the study of valve springs and diesel sprays, in the MIT Sloan Engine Laboratory. A "portable" unit was put together, with help from Kenneth J. Germeshausen, and transported on a truck to Draper's laboratory. The results were immediate and useful. Even today, continuing work is being done in this laboratory with high-speed photography on a variety of technical problems.

We performed a number of experiments with grid-controlled spark gaps. Anyone can make a gap easily, whereas a gas-filled lamp requires a glass blower, pumps, and other back-up equipment. Our first spark-gap lamp was used successfully on a variety of problems. Compared to the mercury-arc lamp, the gap could be put into a specular reflector so that the light could be concentrated on a small subject at a distance. Hardly had the unit been put in a box before we used it for a company that made fiberglass. The results were useful for finding what occurs when a jet of steam is used to blow red hot glass into fibers. This flash unit also produced a loud "bang" because of the explosive nature of the short

---

* bcps originally stood for beam·candle power·second. Today, beam·candela·second is more widely used.

concentrated arc. Unfortunately, I do not recall all the technical details of this "spark" unit. Perhaps the voltage was as high as 10,000 V and the capacity about 10 mF. The energy, $CV^2/2$, was 50 W·s. This energy, released in 10 μs, gave a peak power pulse of 5,000,000 W. That much peak power in the gap for 1/100,000 of a second created a rapid air expansion that could be heard from afar. We put a glass plate over the reflector to help attenuate the acoustical output.

Many of our early photographs were made with this spark-gap equipment. There were those who wanted to make it portable so that it could be used on hand-held cameras. This visionary goal had to wait until the efficient xenon lamp and the lightweight electrolytic capacitor were perfected. Also, improved lenses and photographic films were ultimately very helpful since less output was needed to secure a photograph, making smaller flash equipment adequate.

I remember receiving a phone call from *Life* magazine. "Would your flash unit be adaptable for photographing prize fighters?" they inquired. "Certainly," was my answer. "Let's give it a try." Ken Germeshausen and Herb Grier took a night train to New York, and eventually to New Jersey, to photograph a promising young fighter named Joe Louis. They took with them the high-voltage spark unit, which had a rather low beam output but a 10 μs exposure time. The two boxers got into the training ring and were photographed by the flash unit in a corner near the lamp and the camera. The motion of the fighters was stopped completely by the short exposure time, but the light was far from adequate. Joe's sparring partner had a very black complexion, so his image on film consisted mainly of the whites of his eyes. Louis's image was barely on the film since he was not as dark. The photos were not a big success, but they did give us some valuable experience.

Later, our more efficient xenon lamps were used in an actual prizefight ring. Joe Costa, for example, took an outstanding photograph of the same Joe Louis about to hit one of his adversaries. This picture was published widely. Since then, practically all prizefight photos were made with powerful strobes mounted in the upper part of the boxing ring. As I recall, Costa used three 200 W·s flash units, each with about 8000 bcps output. The lamps were mounted in the ring's superstructure so that the photographer, at ringside, would have adequate lighting, regardless of where in the ring the fight was in progress.

It was significant to me that this first fight photo was published widely and nothing was said about the strobe system that was used. Most previous strobe photographs, when published, had a mention of the system used. This meant to me that strobe (electronic flash) photography finally had become accepted as a standard system of photography.

In 1938, I had a sudden urge to get a high-speed photograph of a football being kicked. At this time, MIT had no collegiate football team, only intergroup, touch football. I called Harvard's athletic department and asked if they would find someone to kick a few balls for me. "Certainly," answered Wesley Fesler, one of the coaches. He said that the boys were all practicing, but he would kick a few. (I found out later that he had had a remarkable football career at Ohio State and was presently on the Harvard coaching staff.)

We met about a half-hour later at one of the covered practice areas near Harvard Stadium. I had brought a double wire synchronizer, which I placed on the far side of the ball, away from the area to be hit by the shoe. When the ball was kicked, the shoe went way into the ball before the ball sprung into action. As soon as the football moved, it caused the wires to contact each other. This completed the trip circuit and flashed the lamp. As I recall, I used the open spark gap lamp, with its exposure time of about 10 to 30 μs. Not only was there a flash of light, but also a resounding "bang" that sounded great. I took three pictures and then packed up my gear and departed for MIT. One of these pictures was a great photo! I have tried several times since to get a better picture, but without success. Only one light was used for the photograph.

About this time I made my first argon lamp. Its efficiency was greater than that of the open spark gap lamp; that is, more light was produced from the same capacitor than from the open spark. Also, it was quieter than the spark since the flash duration was longer and the active lamp volume greater. It seemed to have promise as a photographic light source.

During a public demonstration of this argon lamp in Cambridge, I first became involved with Gjon Mili. We were both scheduled to deliver papers about flash. Mili was first on the program. His experimental mercury-arc lamp blew out the hall's fuses. When I came on the scene, the hall was blacked out. Fortunately, someone fixed the electrical circuits, enabling me to show my slides and demonstrate the strobe lights. Mili was impressed and wanted to know if he could obtain these new lamps. He promised to quit his Westinghouse job and start a studio. He later showed some of his results to *Life* magazine. After this came a series of outstanding results using the strobe system. Mili's story is beautifully recorded in *Photographs and Recollections* (New York Graphic Society, a division of Little, Brown Co., Boston, 1981).

A list of early photographic accomplishments with electronic flash by Mili and others appears in my book, *Electronic Flash, Strobe,* on pp. 125-127. Other examples of this work were published in the MIT alumni magazine, *Technology Review, National Geographic, Life,* and many others. The book *Moments of Vision* (MIT Press) has numerous examples of the applications of electronic flash photography.

Mili was very demanding. We first loaned him a two-light argon flash unit for trial. He said he needed 10 times as much light. Then he said he must have at least five lights to get the proper lighting. So Herb Grier and Kenn Germeshausen got busy. We took a five-lamp strobe assembly to New York with tungsten modeling lamps that had the ability to flash fast and bright. I think Mili was surprised at how quickly we put this assembly together. It was not very portable, however, so other more portable units were made later.

One day George Woodruff, a newspaper photographer for the *Boston American,* came to see me. He wanted a battery-operated portable flash. I tried to get him to say what the specifications should be for the performance he wanted. I had to argue him out of his first, entirely too-powerful unit. I told him that a unit of any size could be built, but that his would be too large and too heavy for typical use. Finally, I got him down to a size that I thought would be reasonable. My criteria was a total weight of about 20 pounds. As I look back on it, this was still too heavy.

For some time we had been trying to interest the Eastman Kodak Company in marketing a flash unit for studio use. Several visits to Boston by Kodak people had been made. We actually had a 200 W·s model, with an output of about 8000 bcps, all built and ready to go. The technical people had suggestions about design changes. It seemed that everyone wanted something different. No one was in a position to freeze the design. After several conferences and much talk, I decided that we were getting nowhere. Our original design seemed fine to us.

I proposed to Woodruff that we take the unit out on news assignments to see if the performance was satisfactory. Every night we took three 200 W·s units to some event, such as a track meet, a boxing match, a skating performance, etc. The negatives were processed and prints were made for publication. Some of the photos were put on the wire transmission service. One in particular was sent to the *St. Louis Post Dispatch* (Feb. 18, 1940) showing four runners in a race. The entire Boston Garden was lit, the depth of field was tremendous, and the *Dispatch* people recognized an outstanding picture and ran it full page. By the end of the week those wire-transmitted photos had created a demand. The sales group at Eastman wanted to get the flash units. Pressure then blew aside all the minor performance nit-picking of the scientists and engineers at the company. Soon an order was received and production was started. We got the General Electric Company in Cleveland to make our strobe lamps and the Raytheon Company in Waltham, Massachusetts, to manufacture the electrical units. The Kodatron was a 200 W·s unit with a diffusing reflector of 18 in. diameter and an output of 8000 bcps. The equipment was designed by Germeshausen, Grier, and the author.

The initial introduction of this unit was made at a photography convention in Chicago (1940). I was present, and my father, Frank Edgerton, came from Aurora, Nebraska, to see the introductions. The Eastman Kodak Company had a professional photographer from Chicago by the name of Kaufman demonstrate the xenon flash units on the stage. He had several models to perform for him. He not only made some portraits, but also showed that action photographs of dancers could be made. The results were outstand-

Back Dive (1964).
© Harold Edgerton.
Courtesy of Palm Press,
Inc.

ing, and the photographic fraternity began to notice. I made an open flash Kodachrome exposure of some of the action.

A nice folder describing the Kodatron unit had been prepared by the Eastman Kodak Company. When I first saw it at the convention in Chicago, I jumped. All through the text it referred to an exposure time of 1/30,000 of a second. Actually, the exposure of the commercial unit was about 1/3000 of a second. Our early spark-gap and some special argon lamps were of a 1/30,000 s duration. The new xenon lamp, with extra energy input and output, was of much longer duration. I called this to the attention of the Kodak group and was told that since the material had already been printed, I should let it go. They said that most of the users of the Kodatron units would not care what the exposure was. It was a mistake on my part to keep quiet. There is no substitute for scientific facts. It was years until the "1/30,000 of a second" comment had been eliminated. Even today, it comes up now and then.

Ever since the early days of flash photography, manufacturers of flash equipment seem reluctant to tell the user what exposure time will result from their equipment. There is no secret about exposure time (flash duration) with electronic flash equipment. For most slow-moving subjects, such as people, there is an inappreciable blur with modern portable or studio strobe lamps. However, those who desire to obtain sharp, clear photographs of bird wings, for example, find that standard equipment will not do the job. Special strobes are required, and information about this is available: for example, see Chaps. 7 and 8 of *Electronic Flash, Strobe,* where details of special equipment for taking bullet and bird photographs are described. A later technical article, entitled "Exposure time: it can be important," in *15th International Congress on High Speed Photography and Photonics,* L. L. Endelman, ed., SPIE Proc. 348, 67 (1982), gives details of a flash unit of 25 μs exposure.

In the early days there were no shutters with a contact that could trigger an electronic flash lamp to flash when the shutter was open. Most of the photos were taken on "bulb" in a darkened room. The flash was arranged to occur when the shutter was open.

Shutters were available with a 20 ms delay for use with the then commonly available chemical flash bulbs. Such a shutter was useless with electric flash since the shutter opened after the flash was all over. Clearly, an instantaneous shutter contact was needed for use with electronic flash. I was amazed at how long it took the shutter manufacturers to develop equipment that was suitable for the new xenon flash. Today, all camera shutters have a contact system that closes a low energy circuit when the shutter is wide open. This contact is called an "X" contact. There is no need for a shutter with a 20 ms lead-time contact since the expendable chemical flash lamp is no longer available.

There are three types of photographers: (1) those who actively use strobe (electronic flash) lighting, (2) those old-timers who still think that "flash" means flash powder or chemical flash lamps, and (3) those who use continuous light, such as sunlight or tungsten light. Actually, the chemical flash lamp suddenly went out of use for several reasons. Those bulbs that burned brightly only once per bulb are now antiques. They are not kept in stock in photographic

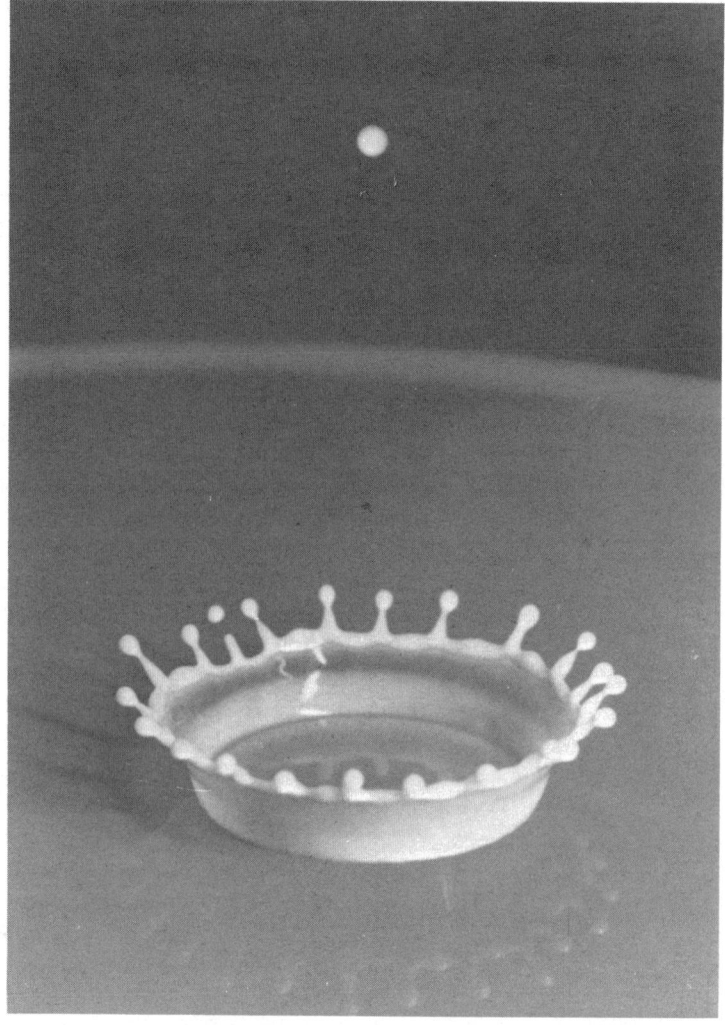

Milk Drop Coronet (1957).
© Harold Edgerton.
Courtesy of Palm Press,
Inc.

stores since the demand has declined to nearly zero. Of course, flash powder has long since become a method of the past. A few of us old-timers still remember those days when a tremendous flash accompanied by a cloud of objectionable smoke resulted for the "one and only" photograph. The smoke pollution made a second exposure impossible.

My first involvement with flash photography was in 1926 or 1927 when I was a graduate student in the electrical engineering department at MIT. I went there to study electrical machinery, especially synchronous motors and generators. A year's experience with the G.E. Company in Schenectady, New York, and lots of other contacts at the University of Nebraska and the electrical light plant at Aurora, Nebraska, were behind me.

At MIT I soon learned that the second-degree differential equation of a motor was nonlinear and, as any mathematician can tell you, does not have a closed-form solution. Fortunately, at that time Dr. Vannevar Bush and his students had developed a calculation machine, called the differential analyzer, that could do the job for specified conditions. With the help of many others, I was able to plot the performance of the machine. It was an exciting time. Would an actual synchronous machine perform like the mathematical solutions? It was in order to check them that I wired up my first strobe. It had to be an improvement over previous strobes because a photographic record was required to record the fast transients of sudden load situations.

I used a grid-controlled mercury-arc rectifier as a light source. An electrical capacitor was used to discharge stored energy into the mercury tube. An internal flash of actinic light of very short duration was produced, which enabled me to take a series of photographs of the angular displacement following a sudden application of load.

The results were very exciting to me and others at MIT. One of those who came to my lab to see the experiment was C. Stark Draper, as mentioned before. He had a series of problems in the MIT Sloan Engine Laboratory for which short exposure photography would be useful. There was a continuous stream of people coming to the strobe lab for technical assistance on all sorts of problems where fast recording was needed.

The mercury-arc lamp was improved for strobe purposes, and eventually a model for use in industry was perfected by the GenRad Company and widely exploited. But, the mercury lamp has a defect. It is dependent on the temperature of the coolest part of the lamp. Therefore, the performance was completely different when the lamp was first started, compared to the performance when hot.

It occurred to us that a noble gas, such a neon or argon, would not be as temperature sensitive. Soon argon was in wide use instead of mercury. Eventually, xenon became preferred because of its better efficiency and spectral distribution.

Today, the xenon-filled flash lamp has taken over almost all applications. Although xenon is very rare (1 part in 100 million), there is plenty of it as a by-product when the oxygen, nitrogen, argon, etc., are removed from air. It was just before the war (around 1937) that several of us realized that electronic flash lighting would be ideal for all sorts of photography, particularly studio photography. Many who looked at the project predicted that it would never take off. They could only see the size and capital expense of the equipment.

Eventually, electronic flash became almost universally accepted due to its remarkable property of efficient production of actinic light of daylight quality. The short exposure time was also important, as was its ability to repeat its flashes.

I recall our initial discussions about the designs for studio use and for portable use. For example, there was a small portrait studio at MIT operated by a Mr. Jackman. He used tungsten lights. One day the wife of one of the professors at MIT came to see me. The photos she had of her superactive young son were blurred. She asked if I could help, and I agreed. I took three lights (2000 W·s) to Jackman's studio and helped him set them up for portraiture. The boy was invited to pose. The results were phenomenal. No matter how much he jumped around, every exposure was sharp and clear. I thought I had made a point, but when I went to see Jackman the following day, he had rolled out the strobe and reinstalled the tungsten lamps. "Why?" I wondered. He had spent a lifetime learning to use tungsten and wanted to continue. I found this reluctance to switch to strobe a common failing of practically all photographers. The solution was to be patient. There was a new generation of photographers on the way who could be taught. Today, practically every studio uses electronic flash lighting for its photography.

Jackman did eventually "see the light" and became one of those who used electronic flash for all of his portraits.

It may be of interest to tell about a request from the Smithsonian Institution for an early electronic flash lamp. I went to Jackman's studio and asked if I could get back one of the lamps after it had been on loan for 20 or 30 years in order to send it to the Smithsonian for an exhibit. He was disturbed until I told him that similar lamps were available in the marketplace.

One Kodatron unit was bought by photographer Charles

Downey of Scottsbluff, Nebraska, who later wrote an enthusiastic story about the new light in his local photo journal. Recently, I met his son, Jim, in Aurora, Nebraska. Naturally, Jim and his son, Tom, take many photographs with a portable xenon flash unit. Jim and his wife, Mona, have appeared in a recent TV tape made by the Nebraska Educational TV Network, Lincoln, Nebraska. The program showed many scenes from the history of strobe photography.

Bibliography

The following references all have extensive bibliographies of articles and books about electronic flash (strobe) photography:

1. H. E. Edgerton and J. R. Killian, *Flash, Seeing the Unseen*, Hale, Cushman and Flint, Boston (1939). Out of print.

2. H. E. Edgerton and J. R. Killian, *Flash, Seeing the Unseen*, Branford Press, Newton Mass. (1954). Out of print.

3. H. E. Edgerton and J. R. Killian, *Moments of Vision*, MIT Press, Cambridge, Mass. (1979). Out of print.

4. H. E. Edgerton and J. R. Killian, *Moments of Vision*, MIT Press, Cambridge, Mass. (1984). Paperback edition.

5. H. E. Edgerton, *Electronic Flash, Strobe*, MIT Press, Cambridge, Mass. (1979); 2nd Edition (1984). Both in paperback.

6. F. Frungel, *High Speed Pulse Technology*, (4 vols.) Academic Press, New York (1965).

Robert W. Gundlach

Retrospective on Xerography and Chester S. Carlson

Robert W. Gundlach joined the Haloid Company in 1952, and participated in the historic growth of xerographic technology as well as the corporate transformation of Haloid to Xerox Corporation. He is a senior research fellow at Xerox. He holds 130 U.S. patents, and received the Xerox Corporation's President's Award in 1979. He was interviewed by Frederick Su, SPIE Technical Consultant, on the invention of xerography by Chester S. Carlson.

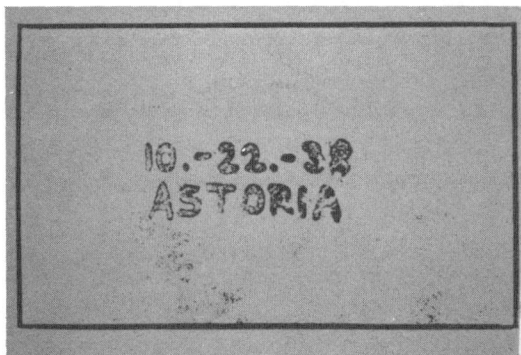

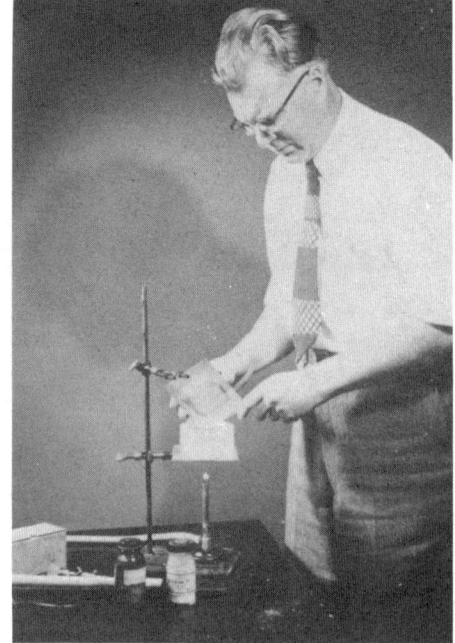

Above: The world's first xerographic image.
Right: Chester Carlson in the 1930s. Photos courtesy of Xerox Corp.

Robert Gundlach

How long has xerography been around?

The first image Carlson ever made actually had the date and the place, Astoria, 10/22/38.

Tell us about the early days.

He probably had the basic idea in 1936 or '37. He was living on Long Island. He did work in his kitchen, melting sulfur onto a zinc plate. He would put the sulfur powder on a pan and warm it over his gas stove. It would melt and he would tip it so it would flow uniformly over a three- or four-inch square plate. Several times it caught fire and that made his experiments pretty unpopular with his wife; so he rented a room behind a beauty parlor in Astoria, Long Island. He worked alone there for quite a while. Finally, realizing he needed help, he hired Otto Kornei. This was around October 1st. Within three weeks they succeeded in making the first image. It was crude compared to now, of course. They rubbed a sulfur-coated plate with a handkerchief to get it charged. On a little glass slide they had written in India ink, 10/22/38, Astoria. Then they placed the slide, image side down, against the sulfur plate and exposed it to a 100-watt light bulb, which was about eight or ten inches away. They exposed the slide and plate to the light for about eight seconds. I've seen his record of this in his notebook.

The concept is to use a photoconducting film on a conducting substrate such as a metal plate. The photoconducting film, sulfur in this case, will hold charge in the dark and lose charge in the presence of light. After exposing to a pattern of light, you have an electrostatic latent image that can be developed with charged particles of powder.

Did they rub the zinc plate or did they put the sulfur on top of the plate and then rub it to charge it?

The sulfur was put on the plate first so that it was essentially a coated plate (like a coating of paint). They would rub the coating to charge it. Then they would selectively discharge with a pattern of light so that a pattern of electrostatic charges remains on the film and can now be developed with electrostatically charged powders. The image is made visible just by pouring fine powder over it. The powder is like kitchen flour, but dyed black. Actually, that first image was made with dyed lycopodium powder; the spores of star mass. In the first reduction to practice 50 years ago, Carlson transferred the image to wax paper by pressure; actually he rolled wax paper over the powder developed plate.

How has the process changed over the years?

That first reduction to practice was very crude. The sulfur had low sensitivity to light so they handled it in a dimly lit room. It would have taken several hours for the latent image to form on the plate in a camera. If Batelle hadn't come up with selenium some years later, it wouldn't have been a viable process. But it did demonstrate the feasibility of the concept.

In the earlier days, the photoconducting films had to be highly insulating in the dark, so they could hold charge for at least a minute when the process was done manually. Today's photoreceptors could decay in 10 or 20 seconds and still would be workable because the drums go so fast that you can form the latent image and develop it within three or four seconds, sometimes within one or two seconds. So you don't really need low-dark-decay photoconductors in the high-speed machines.

The charging is better today also. Almost all of the machines have the photoreceptor drums or belts charged coronally; i.e., ions of air are sprayed onto the surface of the drum. This was suggested by Carlson and developed by Battelle Memorial Institute in Columbus, Ohio. They also invented electrostatic transfer, so you could transfer the image to plain paper instead of wax paper. Electrostatic transfer is done by placing plain paper against the powder-developed image and corona charging the back of the paper with a polarity opposite the charge on the powder, so that the powder is attracted to the paper.

Okay, then how do you keep the powder on the page without it rubbing off?

The powder is called thermoplastic, meaning it melts when you heat it. Think of the powder as a pulverized wax. If you had a powder of wax, once it is on the paper, just elevating the temperature would convert it to an ink. The ink is then absorbed by capillary penetration into the fibers of the paper. So a xerographic image—those copies that you see on your papers on your desk—are thermoplastic powder images melted into the paper. And they are as permanent as the paper itself.

I was surprised that this invention was so long ago. That he saw the need for office reproduction in the 1930s!

Isn't it amazing? And he couldn't talk anybody else into it. Even after he showed that it worked, they didn't believe that it would be commercially viable, that there would be enough commercial interest to warrant licensing it for development. It took Carlson six years to get anybody interested at all. The Battelle Memorial Institute heard about it and agreed to sponsor work on it if he would put up some money himself. I think he had to borrow $15,000 from his cousin.

When did xerography first catch on? Did Carlson give the patent to Xerox Corporation?

Carlson licensed the process to Battelle, and they made very substantial improvements on it. After several years, someone at Haloid Company read about the process and they negotiated with Battelle for license and co-development with Battelle. Haloid received all the marketing rights for xerographic copiers.

What year was this?

In 1948. Actually, in late 1947 they were negotiating it. I think it was finally legally drawn up and completed in January 1948. The basic process was first presented to the world at an Optical Society meeting in Detroit on October 22, 1948—10 years to the day after the world's first xerographic image.

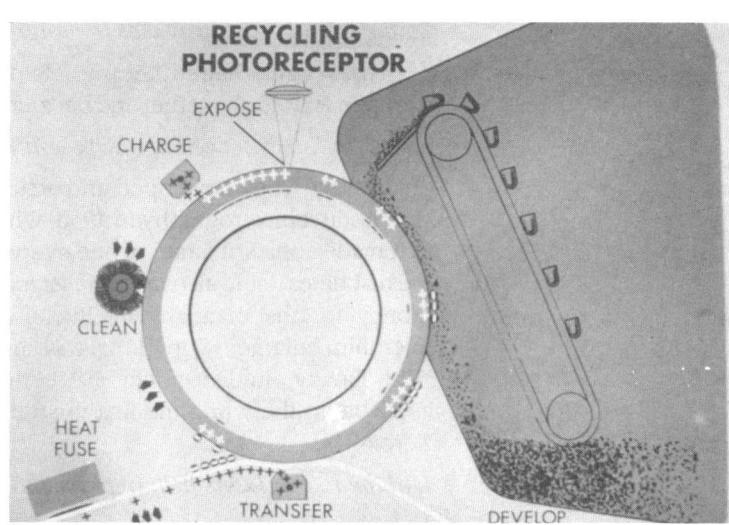

Process steps of the Xerox 914—the basic steps for plain-paper xerography. The photoconducting drum is charged with a corona unit. The charged region then passes under a synchronously moving projected image, which selectively exposes and discharges the drum, creating the latent image. The electrostatic latent image is then developed with charged toner particles and the developed image is transferred to plain paper by spraying the paper with charges opposite in polarity to the toner charge. The toner particles are then transferred to the plain paper as an image and this image is then melted into the paper by heating at the fusing station. Residual toner is brushed clean from the drum, and the drum is then ready for recycling the next copy.

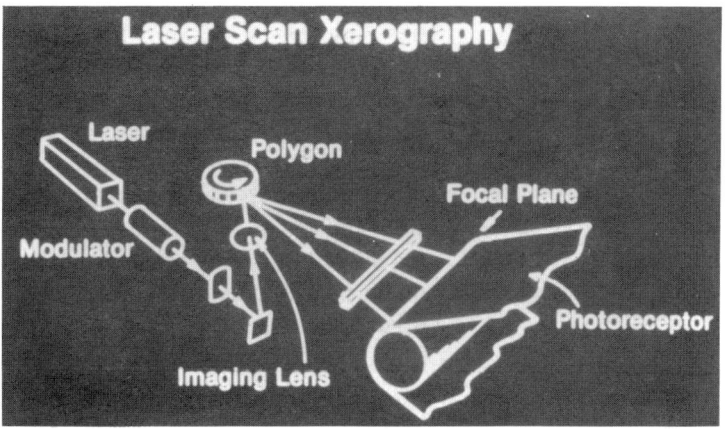

Laser xerography—the optical components for scanning a laser beam across a photoreceptor. The beam is sent through a modulator to switch it on or off on command. From there it is reflected to a multifaceted polygonal mirror spinning at high speed that sweeps the light beam across the belt or drum. A cylindrical lens is shown just before the photoreceptor. Its purpose is to reduce the effects due to wobble in the spinning polygonal mirror.

So the Haloid Company was the forerunner of Xerox?

That's right. Haloid became Haloid-Xerox in 1958 and then dropped Haloid altogether and became Xerox Corporation in 1961.

So in 1948 they picked up the patent rights.

Correct. They developed and actually made the first commercial product in xerography in 1950, which was a flat plate machine that made copies at a rate of one every three minutes. Fortunately, the machines could also make offset masters, and that is what made money for the company the first five years. Then we built a microfilm enlarger (Copyflo-II) that made some money in the late '50s. Finally, in December 1959, they announced and made available to the world the first push-button automatic copier, the Xerox 914.

Did the role of lasers help improve the process?

Yes. After 15 years of phenomenal success in copying, the usefulness of the xerographic process was extended to electronic printing, by exposing with a scanning laser beam instead of a camera lens. Rather than having to go from an existing document through a lens in a camera to project the visible image onto the plate, images electronically created by computers or word processors can be written directly onto the photoreceptor with a pencil of light, scanning much like the electronic beam scans the image onto your TV screen. When a xerographic drum is totally charged, you can selectively discharge it with this pencil of light that writes across the photoreceptor a line at a time. It is what they call bit mapping. Let's say a line of characters can be 30 scan lines high, so as you write one scan-line across the photoreceptor with the laser beam you will see that the line will have to go dark in some spots and light in other spots. These little pixels, then, will be white or black. We can do this at 200 scan lines per inch. Actually, we are doing 400 lines per inch now. So you have one line .0025 of an inch wide and you turn the laser beam on or off at the right points (about a million times every second) and you go just one line down and do the next, and so on. If you go line by line, you keep adding them up and pretty soon you have a whole line of characters. So that is how your latent image is formed on the drum.

Why is electronic imaging better than conventional imaging using a camera?

In terms of image density and clear background it can be better because the lens projects smudges of dirt or weak (light gray) images, whereas electronic images are represented by binary code; i.e., each pixel is either dense black or clean white. It doesn't have to deal with various shades of gray. Continuous tone pictures can be simulated by writing halftone dots, much as the printing industry has done for generations.

In terms of flexibility, electronic images can be transmitted all around the world and transformed back to hard copy. Even image size and contrast can be changed easily, once it has been digitized.

When did the Carlson patent expire? Because now everybody uses this process.

The basic patent was issued in 1942.

What is the usual run?

Seventeen years.

That's all?

In seventeen years we hadn't even introduced the first copier yet. Let's see, 42 and 17 is 59. We introduced the copier the year the basic patent expired. But the patents on corona charging were still in force, as well as the patent on electrostatic transfer. So nobody could really practice a commercially viable process with plain paper until those patents expired (in 1968, I believe).

Did Carlson reap any monetary benefits from this?

Yes. It should be said that that wasn't his goal, although he was driven by the fact that his family was very poor and he had wanted to make some money. But it was clear that after he made it he felt an obligation to see that it was used well. He didn't live an extravagant life at all. He went to Europe once with his cousin to see a show on xerography and he traveled second class on a steamer.

Xerox did pay him for the use?

Yes, but not as an employee. He was a consultant for Xerox, but to my knowledge he was never on salary with us. It was just because of his royalty rights to the process that he served as a consultant. We gave him a desk, and he came into our labs, and worked in the patent department too. But Carlson did reap about $150–200 million based on royalties.

That's not too bad.

Right. Some of that he had to give to his cousin who had advanced the $15,000. He gave him a fraction of the royalties.

That's a good investment.

That's right, but nobody else thought so at the time. In fact, some companies who sued us later for rights to use the process were suing on the basis that we were monopolizing the process.

> *"Nobody wanted to help bake the cake, but they all wanted to taste the frosting."*

Of course, anybody could have invested in these further patents, but they chose not to. In fact we asked them in the '50s to help and they didn't. As our lawyers said later, "Nobody wanted to help bake the cake, but they all wanted to taste the frosting."

Discovery and Innovation

But one unclouded blaze of living light.
Lord Byron, *Curse of Minerva*

Willis E. Lamb, Jr.

Einstein and Laser Theory

Willis E. Lamb, Jr. received the PhD from the University of California, Berkeley, in 1938. He was a professor of physics at Columbia University, Stanford University, Oxford University, Yale University, and a professor of physics and optical sciences at the University of Arizona. He received the Nobel Prize in physics in 1955. He is a Fellow of the American Physical Society, a member of the National Academy of Sciences, an Honorary Fellow of both the Royal Society of Edinburgh and the Institute of Physics and Physical Society, and an Honorary Life Member of the New York Academy of Sciences.

The conventional view is that it all began with the wonderful 1917 paper of Albert Einstein, who introduced the famous A and B coefficients and deduced the necessity for the new process of stimulated emission radiation. He was studying the conditions which must be met in statistical mechanics in order to get the Planck distribution law for radiation in a black body cavity in thermodynamic equilibrium. Einstein considered what would happen to atoms placed in the radiation field of the cavity. Only two stationary states of these atoms came into the discussion. He introduced the A coefficient to describe the rate of spontaneous emission of radiation by an atom, initially in the upper state, making a transition to the lower state. He also postulated that an atom, initially in the lower state, could absorb radiation at a rate given by a coefficient B multiplied by the appropriate (for resonance) spectral density of radiant energy in the holhraum. In order to get the correct equilibrium distributions for radiation and matter, he found that he also had to assume that there was stimulated decay whereby an atom in an excited state would, in addition to the spontaneous decay, also make stimulated radiative transitions to the lower state at a rate given by the product of the same constant B and spectral density of radiation.

After the development of the quantum theory of radiation by Dirac in 1927, it became clear that one's ideas needed to be somewhat rearranged. One only needed the quantum mechanics of 1925-1926 to describe how a classical electromagnetic field could cause stimulated (or induced) transitions of an atom in either direction. Of course, one thereby only saw that the atom was changing state, and not that radiation was being absorbed or

© 1984 IEEE. Excerpted with permission from "Laser Theory and Doppler Effects," *IEEE Journal of Quantum Electronics*, Vol. QE-20, No. 6, p. 554, June 1984.

emitted. An energy conservation argument could be used to fill the gap in the physics. The quantum theory of radiation was needed to deal with spontaneous emission and, of course, led to a proper description of the stimulated processes as well. That theory also made possible a rational discussion of problems involving light quanta or "photons," but the quantum optics community has been slow to appreciate this fact. See [1, pp.98-104] for more comment of this point.

It's easy to see, in retrospect, why Einstein took spontaneous decay as a very fundamental process. He had pretty well rejected the Maxwellian electromagnetic theory. He knew that there was spontaneous radioactive decay of nuclei leading to alpha particle emission, and also of the spontaneous emission of radiation involving transitions from an upper to a lower stationary state as described by the Bohr theory of 1913. It is a little harder to guess Einstein's reasons for making use of absorption in his argument. He may have simply taken it as experimentally well established that light is absorbed in matter.

We can get a simple view of absorption by thinking of a test charge q placed in an electrostatic field E. The charge will have a force qE on it in the direction of the field, and if its initial velocity v_o is also in that direction, the velocity v and the kinetic energy $mv^2/2$ of the charge will increase in time. If the electrostatic field were a prescribed and unchangeable entity, one would have to say that the increase of kinetic energy was associated with a decrease of "potential" energy. Presumably, the discussion could be extended slightly to allow for slowly varying electromagnetic fields. If these Maxwellian fields were treated dynamically, and the techniques for doing this were available long before 1917, there would be an energy associated with the electromagnetic field, and any gain in the kinetic energy of the particle would be accompanied by a decrease in the energy of the electromagnetic field. We thereby have a model for *absorption* of radiation. If the charge were initially moving in the opposite direction to the electric field, the Lorentz force would slow it down and, at least for a certain span of time, it would lose kinetic energy which would doubtless find its way into the energy of the electromagnetic field. Here we have the simplest form of stimulated emission. Whether we have absorption or stimulated emission is only a matter of the relative phases of the motion of the particle and the electromagnetic field. Some counterpart of spontaneous emission can also be found in such classical discussions, but as discussed in [1, pp. 105-110] the modeling is not nearly so faithful as it is with classical models for the stimulated processes. Einstein did not have any adequate dynamical theory of his quantum of light, but it seems safe to say that whatever theory he could have had would have treated absorption and stimulated emission on an equal basis. Then he could have deduced the value of his A coefficient from the validity of the Boltzmann and Planck distribution laws 10 years earlier than the time when the quantum theory of radiation could be used to make the calculation.

Reference

1. W.E. Lamb, Jr., "Physical concepts in the development of the maser and laser," in *Impact of Basic Research on Technology*, B. Kursunoglu and A. Perimutter, Eds. New York: Plenum, 1973, pp. 59-111.

Charles H. Townes

Ideas and Stumbling Blocks in Quantum Electronics

Quantum electronics, particularly the maser and the laser, represents a marriage between quantum physics and electrical engineering which was probably longer delayed than it might have been because the two were not sufficiently acquainted. In this article, Nobel laureate Charles Townes discusses the mutual discovery of one field by the other, as well as the misunderstanding and false starts encountered along the way. Specific examples are used to make more real the thinking of the early years in this field and the struggles with ideas which, as with most now-understood sciences or technologies, seem much simpler in retrospect.

It is sometimes said that there is no single component idea involved in the construction of masers or lasers which had not been known for at least 20 years before the advent of these devices. Of course, a discovery which might have occurred earlier is not uncommon in science or engineering. Nevertheless, the case of quantum electronics is striking enough that it may be useful to review the development of ideas prior to the time this new field became visible, and the stumbling blocks which may have delayed its creation. I believe whatever unnecessary delay occurred was in part because quantum electronics lies between two fields, physics and electrical engineering. In spite of the closeness of these two fields, the necessary quantum mechanical ideas were generally not known or appreciated by electrical engineers, while physicists who understood well the needed aspects of quantum mechanics were often not adequately acquainted with pertinent ideas of electrical engineering. Furthermore, physicists were somewhat diverted by an emphasis in the world of physics on the photon properties of light rather than its coherent aspects. It is still surprising that the basic combinations of ideas required for quantum electronics were not more completely envisaged somewhat earlier than they were. Nevertheless, it is understandable that the real growth of this field came shortly after the burst of activity in radio and microwave spectroscopy immediately after World War II, since this brought many physicists into the borderland area between quantum mechanics and electrical engineering.

I know that most of my electrical engineering friends, while well acquainted with absorption of radiation by atoms and molecules, were surprised to learn that excited molecules could give up energy coherently to an electromagnetic wave. With that knowledge in mind, they might at least have imagined utilizing such effects for amplification even if they were not expert with the specific arrangements required. Many physicists knew of stimulated emission, but few connected it with useful amplification. That amplification by stimulated emission had been well understood by many individuals is impressively demonstrated by a number of early records—from Richard Tolman's publication in 1924 discussing amplification by an inversion of population, to serious consideration of an experiment to demonstrate stimulated radio emission by John Trischka at Columbia University several years before invention of the maser. In all, I know of eight apparently independent discussions prior to the maser invention of how stimulated emission and nonequilibrium populations can increase the intensity of a wave. There may be others.

Why were the many discussions of stimulated emission not followed up to produce some actual demonstration or a useful device? In some cases such effects appeared to be only of theoretical interest, providing a neat and consistent explanation for characteristics of radiation and of absorption. Furthermore, the simple weakening of absorption with increasing population in an upper state (e.g., for $h\nu > kT$) always seemed to me an adequate demonstration of stimulated emission. The few experiments on stimulated emission attempted in the optical region made demonstration of any large effects seem difficult, and were not followed up by other experimenters. Probably the failure to couple the idea of feedback with weak stimulated emission helped make such effects seem inevitably small. In the case of Trischka, and in my own thinking before a feedback oscillator was envisaged, a demonstration of amplification in the microwave region due to inversion seemed rather difficult and not important enough to be worth the required time. In the case of inversion of nuclear spins, which was obtained some years before maser action, the low frequencies and nuclear magnetic dipole moments involved made most stimulated emission effects rather small, and presumably for that reason successful inversion of populations did not direct attention toward useful amplification.

In my own case, it was primarily a strong desire to obtain oscillators at shorter wavelengths than those otherwise available that induced me to initiate experimental work on the maser. This is why the first system designed on paper was for 1/2 mm wavelengths, in the far infrared. My students and I had previously tried many techniques—magnetron harmonics, coherent Cerenkov radiation, and others to obtain short wavelengths, and while most of them worked after a fashion, none gave the promise which masers did for good spectroscopic sources at short wavelength. Initially, I did not fully realize the maser's potential as a low-noise amplifier or as a precision clock, though these two applications added considerable interest rather soon after work on a maser was started with James Gordon and Herbert Zeiger.

As indicated above, my belief is that for many of the physicists who understand stimulated emission, isolating such effects seemed somewhat difficult, and the necessary experiments were not very important because to these same physicists stimulated emission was already rather well understood. The idea of feedback and large numbers of quanta in single modes which might have suggested practical applications and given additional value to such experiments had not occurred. In addition, there were some misunderstandings and confusions which played a role in delaying quantum electronics. About 1945, I had myself written an internal memorandum at the Bell Telephone Labs explaining that molecules and atoms could be used to generate short microwaves, but that intensities could only be low because they would be limited by the second law of thermodynamics—use of nonequilibrium distribution and population inversion had not yet occurred to me.

Emphasis on the photon aspect of light deflected some physicists from coherent amplification. It turned out that before the maser was operational, John Von Neumann had suggested exciting electrons in a solid by neutron bombardment and thereby obtaining a powerful cascade of photon emission. Coherence was not mentioned, and I believe no one attempted such an experiment. J.H.D. Jensen told me that in the 1930s he had thought about stimulated emission from an inverted population as a cascade of independent photons, like a cosmic-ray shower. He lost any great interest in the idea after an experiment which seemed to produce such effects turned out to be explained otherwise. In thinking about light itself, rather than microwaves, it may be that many electrical engineers would have not been any more concerned with coherence effects than were these physicists. However, both engineers and physicists were naturally led to consider coherence when dealing with the radio or microwave region, and I believe this is why initial ideas and development of the field were so dependent on those with experience in radio and microwave spectroscopy.

Some of the confusion about coherence seemed a little strange even in the early days, and will seem remarkable at this time, but was real. Consider the early experiment of A.T. Forrester. Shortly after World War II, he began an experiment to irradiate a photoelectric surface with two Zeeman components of an optical line in order to mix the two frequencies. The idea was to have the two frequency components separated just enough to produce a varying photoelectric current at a microwave frequency low enough to detect by available electronics. Forrester's interest, I understood, was to demonstrate the effect of mixing two discrete optical lines, an effect which should have been detectable by the recently developed high-frequency electronics. The experiment was not easy at the time, and apparently a number of physicists believed it to be conceptually wrong. There seemed to be confusion over the spatial extent of the possible coherence, and questions whether the independent particles of different frequencies could cooperate in emitting electrons from a surface and thus give a beat. My belief then and now is that a bright electrical engineer would have figured out that the experiment would work. Nevertheless, the basic idea was challenged by a publication in the *Physical Review* in 1948 and by enough scientists that, after the experiment was under way, I was asked by the sponsoring agency to review it and advise whether the idea was faulty. The experiment was only difficult, not erroneously planned, and Forrester's published result later dispelled any doubts.

Perhaps a somewhat more subtle example of physicists' bent at the time toward thinking in terms of individual particles involved the frequency spread of a maser oscillator. From the point of view of stimulated emission produced by an oscillating field established in a resonant cavity, it is not hard to understand that the radiation produced by a maser oscillator could indeed have a very narrow frequency band, independent of the width of response of the individual, excited molecules. Any real width has to be due to either the small amount of additional spontaneous emission or to thermal radiation present, as first worked out by James Gordon. However, there was the uncertainty principle relating time and energy, a basic law for physicists. With the lifetime t of molecules in the cavity limited (for the beam-type maser) by the time of transit, how could there be a frequency width much smaller than $1/t$? An electrical engineer accustomed to the almost monochromatic oscillation produced by an electron tube with positive feedback would perhaps not have given the problem a second thought. However, before oscillation was achieved I never succeeded in convincing two of my Columbia University colleagues, even after long discussion, that the frequency width could be very narrow. One insisted on betting me a bottle of Scotch that it would not. After successful oscillation, I remember interesting discussions on this point with Niels Bohr and with Von Neumann. Each immediately questioned how such a narrow frequency could be allowed by the uncertainty principle. I was never sure that Bohr's immediate acceptance of my explanation based on a collection of molecules rather than a single one was because he was convinced, or was due simply to his kindness to a young scientist. My discussion with Von Neumann had a more special twist and occurred at a social occasion. When I told him about our maser oscillator, he doubted that the uncertainty principle allowed our observation of such a narrow frequency width to be real. After disappearing in the crowd for about 15 minutes, he came back to tell me he now understood the situation; my argument was correct. That Von Neumann took even that long to understand impressed me. He then went on to urge that I try for stimulated emission effects in the infrared region by exciting electrons in semiconductors. I was puzzled by his strong insistence on this, because at that time the use of semiconductors for stimulated emission amplifiers seemed much more difficult than other possible methods. It was only much later I learned that he had already independently proposed such a system to produce a powerful avalanche of photons by stimulated emission, and must have been avoiding the more usual response by saying he had invented the idea before he heard of our work.

I must note that, although a number of good physicists did not find the coherent aspects of masers straightforward, for most of those in the field of radio and microwave spectroscopy, they were fairly obvious. That was true, for example, of my colleagues I.I. Rabi, Polycarp Kusch, and Willis Lamb, and of course Arthur Schawlow with whom I later collaborated on the laser. Electrical engineers, while not so knowledgeable about quantum properties, also found coherence properties very natural and seemed to have the right instincts about many aspects of quantum electronics from parallels in ordinary amplifiers and circuits. As an illustration, I give an example of how an electrical engineer helped me at one point.

The frequency stability of a maser oscillator was an important question, and I had worked out an expression for it which showed, I thought, that the cavity pulling would give an error proportional to the square of the ratio of quality factor Q for the spectral line to that of the maser cavity. On explaining this to an electrical engineer at Stanford (whose name unfortunately I cannot recall), he remarked that such a result was peculiar since frequency pulling of one resonance by another in circuits was proportional to the first power of this ratio. While denying any detailed knowledge of the quantum mechanical properties, he doubted my result was correct and he was right. A little further examination when I returned home showed that I had neglected reactive terms of the quantum mechanical oscillator and cavity, which then gave just the result he expected.

As of today, the more engineering ideas of coherence, feedback, and nonlinear frequency mixing have become so intermingled with the more physical ideas of discrete states and quantum mechanical processes in the minds of both electrical engineers and physical scientists that some of the above confusions will probably be hard to believe. That is why the few specific examples above may help remind us how things were.

In addition to conceptual stumbling blocks which affected the course of quantum electronics, in the early days there was also a limited appreciation of the potential of this field, and this too may deserve illustration. Of course, I do not pretend to have foreseen the field's full potential myself, though I was obviously more impressed by it than many others.

Consistent with my usual practice of working with graduate students, development of the maser proceeded at the rate of a normal graduate thesis project, being completed by Jim Gordon approximately three years after the idea was conceived. Our laboratories at Columbia University were completely open in the usual way of academic institutions, and many knowledgeable people visited the maser experiment. However, no one seems to have thought it interesting enough to reproduce or to try to compete with us. As far as I know, there were no concurrent efforts to obtain a molecular or atomic oscillator except the work of Basov and Prokhorov in the Soviet Union, and in this case I am not familiar with just what was done in these earliest years.

Completion of the maser oscillator was exciting to some, but evoked no more than mild interest on the part of other of my friends and did not immediately generate any great flurry of work. For some time it appears that the potential of quantum electronics was unappreciated by many of those not already in the field. In part this result could have been because the first maser itself may have been judged both limited and specialized. But also, for understandable reasons, scientists busy with their own research are not necessarily quick to see the potential of new events in other fields. Whatever the reason, when the performance of masers as frequency standards and then as amplifiers became more evident and as new varieties such as many solid-state systems were proposed, interest in masers grew. By 1960, publications on new varieties of masers became so common that, presumably under the assumption that the excitement must be over, *The Physical Review* made public a policy not to accept any more letters on new masers.

Completion of the maser oscillator was exciting to some, but evoked no more than mild interest on the part of other of my friends and did not immediately generate any great flurry of work. For some time it appears that the potential of quantum electronics was unappreciated.

The delay of about six years between masers in the microwave region and lasers, or masers at shorter wavelengths, was no doubt also due in part to some conceptual stumbling blocks. One of these was imperfect recognition of the possibility of obtaining a high Q and of emphasizing a single mode in a structure which is very large compared to a wavelength, like the Fabry-Perot, even though Fabry-Perot resonators were well known. This problem and some others made it difficult to recognize ways that masers operating at infrared or optical wavelengths could perhaps be as easy or easier, rather than harder, than those in the microwave region. I shall not here try to explore the missing conceptual links, but rather turn to a different aspect, an apparent lack of appreciation of the potential of optical masers prior to late 1957. A number of individuals certainly recognized that maser techniques might be extended to much shorter wavelengths. I believe it was in 1956 that Bill Otting, Head of Physics for the Air Force Office of Scientific Research, asked me if his office could support me or someone else I might suggest in work toward an infrared maser oscillator. It is difficult to remember how many other more casual conversations there might have been on the subject or to know how many scientists may have considered this, but there were apparently no substantial efforts to explore maser oscillators at wavelengths much shorter than the microwave region before 1957. I know why I myself delayed this long—I was busy with and excited by microwave applications of the maser and saw only rather brute-force methods of moving to much shorter waves before that time. I was waiting for a "neater" idea to occur. About others I have no direct evidence, but believe a lack of appreciation of the potential of lasers and a closely connected effect due to the state of development of optics and optical oscillators both played a role in the time-delay between microwave and optical oscillators.

Optical spectroscopy had its heyday for physicists in the 1920s and 1930s; by 1940 most physicists considered it a mature field of solid importance but from which no remarkable breakthroughs were likely to emerge. There was, I believe, an attitude of déjà vu about optics. After World War II optics and optical spectroscopy did have a substantial renaissance, especially in the hands of French physicists, but was still not an area to which many turned for forefront physics. As I see it, lasers might well have been invented during this 1925-40 period, although they would have been more difficult for lack of certain techniques which were further developed in later decades. These include a miscellany of things such as good optical coatings, flash tubes, and improved varieties of infrared materials and detectors.

As the possibility of high-quality, single or almost single-mode lasers came into everyone's view, interest and intensity of attention to this field increased sharply. Nevertheless, much of its now quite evident potential was initially appreciated only by enthusiasts, and some, not even by them. There was almost immediate interest in the optical maser proposals of Schawlow and myself, but the beauty of the device may have been more attractive to most scientists than its potential applications. A favorite quip which many will remember was "the laser is a solution looking for a problem." While an enthusiast myself, and aware of the potential for high-precision measurements, monochromacity, directivity, and the high concentration of energy that optical masers would provide, I missed many potent aspects. The area of medical applications is one that did not occur to me initially as promising. In retrospect, I can imagine recognizing the beauty of operating directly through the pupil without other insults to the eye, but since I had never heard of a detached retina, such an idea would have been another "solution looking for a problem." My own scientific interests were primarily in the direction of new forms of spectroscopy and precision measurement, and hence I needed only modest power. While it was evident that optical masers could be expected to produce powers of at least a number of watts, I did not initially think of very short pulsed operation at a power level of many kilowatts, as was produced by Maiman's ruby laser.

In looking back over why the field of quantum electronics took as long as it did in getting started and why even then the buildup was initially not more rapid, I necessarily mention some of the stumbling blocks, misconceptions, and fumbles. The development of any science by humans has its similar mistakes and illogicalities. Recalling these can keep us humble and make us aware there may be other exciting events not yet visible around the corner. However, focusing on problems of the past omits or de-emphasizes the remarkable insights and inventions made by a large number of colleagues who have contributed to this field, and the vigor with which industry pursued and developed it. I can resist discussing these impressive aspects of the field only because I know others will treat them appropriately.

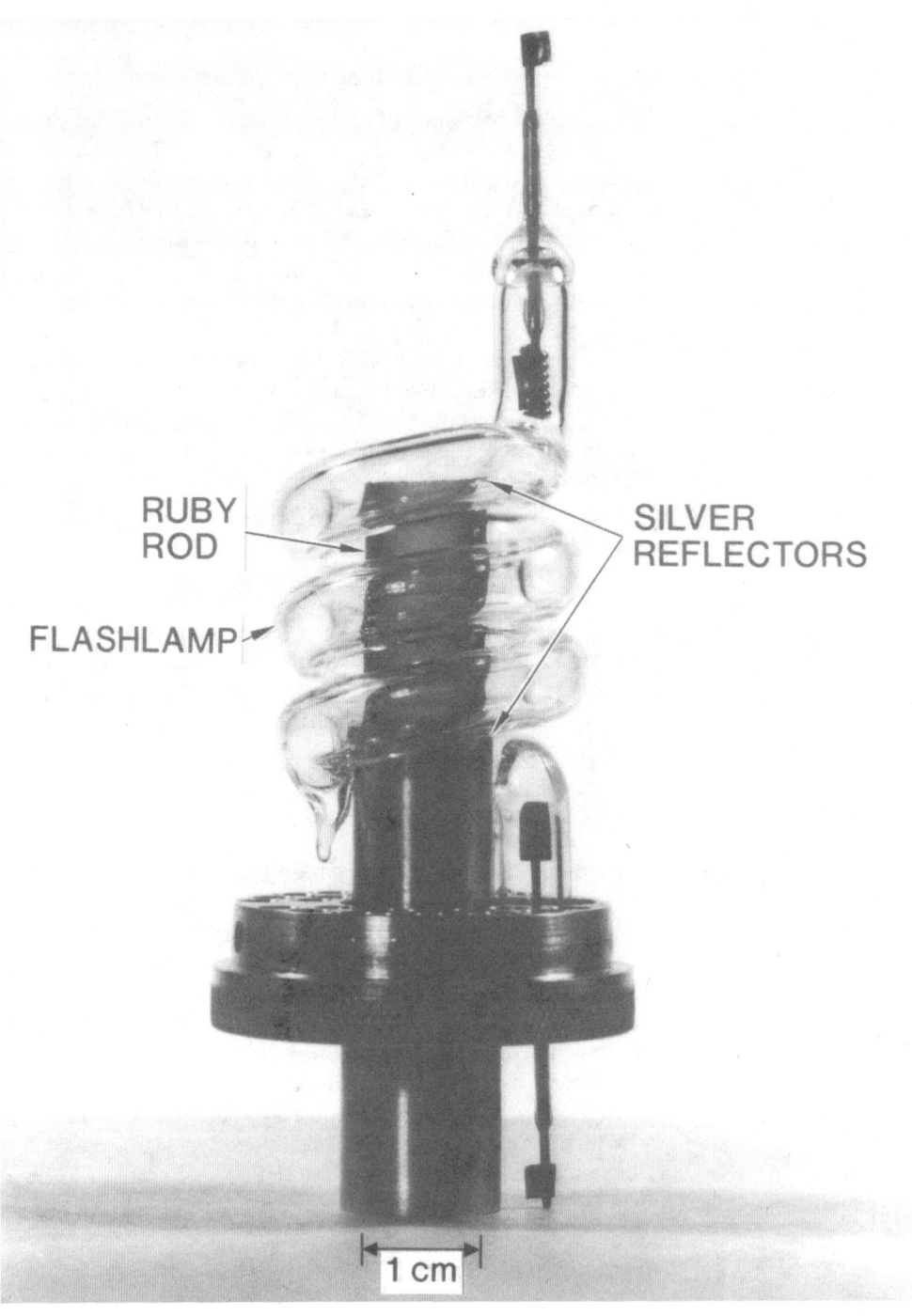

RUBY ROD

SILVER REFLECTORS

FLASHLAMP

1 cm

The world's first laser. Theodore Maiman used this ruby laser to produce coherent optical radiation on May 15, 1960. Photo courtesy of Hughes Aircraft Company.

George F. Smith

The Ruby Laser

George F. Smith received the PhD from California Institute of Technology in 1952. In 1960-61, he identified the rangefinding potentialities of the ruby laser, and conducted the first laser rangefinding experiments at Hughes Research Laboratories. He is a Fellow of the American Physical Society.

The story of the ruby laser has its roots in T.H. Maiman's ruby maser program, which commenced in 1957. At that time, the new ruby maser was being investigated at several laboratories. Maiman succeeded in building a cavity-type maser that was lightweight, compact, and generally suitable for use as a low-noise, X-band preamplifier for communications or radar systems. When operated at 4.2 K, it gave a 100 MHz gain-bandwidth product, and a noise temperature of 2.4 K. At a more practical operating temperature of 77 K, the noise temperature was still a respectable 93 K.

Early in 1959, Maiman examined the possibilities of optical pumping of the maser. Since the theoretical expression for the noise temperature of a maser amplifier varies as the reciprocal of the pump frequency, it is obvious that a much lower noise temperature could be achieved if the pump frequency were increased from the conventional 24 GHz to an optical frequency. Wieder, at Westinghouse, also was exploring optical pumping. He tried to pump a ruby maser with the fluorescence from a second cooled ruby. Wieder [1] attributed the failure of his experiment to a low value (1 percent) for the quantum efficiency of ruby fluorescence.

Meanwhile, the classical laser article by Schawlow and Townes [2] appeared in December 1958. This article described generally how one might build a laser, and listed some of the problems to be solved. During 1959, interest in the possibility of making a laser grew rapidly at many laboratories, in America and abroad. Maiman felt that a solid laser offered some advantages: 1) the relatively simple spectroscopy made the analysis tractable, and 2) construction of a practical device should be simple. Although Maiman had at first given some consideration to the use of the gadolinium ion, in gadolinium methyl sulfate or some other salt, he was drawn to the possibility of making a ruby laser, perhaps due to his extensive experience with the ruby system. In fact, there were two counts

© 1984 IEEE. Excerpted with permission, from "The Early Laser Years at Hughes Aircraft Company," *IEEE Journal of Quantum Electronics*, Vol. QE-20, No. 6, pp. 577-579, June 1984

against ruby: 1) the literature value for the quantum efficiency of ruby fluorescence was very low, as mentioned above, and 2) the fluorescent transition upon which the prospective laser was based terminated in the ground state. The second factor, especially, had led many experimenters [3] to rule out ruby as a laser candidate since it meant that the achievement of inversion would require the emptying of more than half of the population of the ground state. Lasers in which the laser transition terminates on an excited state do not have this constraint.

At this point, Maiman decided to remeasure the quantum efficiency of ruby fluorescence. The prospects for both the optically pumped ruby maser and the ruby maser depended critically on that parameter. Furthermore, Maiman felt that it was important to understand why the quantum efficiency might be low, to guide a search for a better material, should that be necessary. When Maiman's measurements [4], [5] gave a value of 70 percent for the quantum efficiency, the race was on!

First, Maiman developed criteria [6] for laser action as a function of gain per pass and mirror reflectivity. He concluded that the brightest continuous lamp readily available, a high-pressure mercury vapor arc lamp, would be marginal. A pulsed xenon flashlamp, on the other hand, appeared promising.

Before proceeding with the laser experiment, Maiman conducted an optical/microwave experiment [4], in which he monitored the 11.3 GHz absorption of a cube of ruby during application of a pulse of light from a xenon flashlamp through a Lucite light pipe. A significant decrease in the microwave absorption indicated that an appreciable fraction of the ground state was being depleted. With this encouragement, Maiman was ready to build his ruby laser.

As it happened, the timing was poor. In the spring of 1960, just when Maiman was ready to start his laser experiments, the Hughes Research Laboratories commenced their move from the Hughes Culver City plant to their new (present) home at Malibu. Needless to say, Maiman's group made the move very expeditiously; he was back in the lab within two weeks, during which time he prepared the paper [4] reporting the quantum efficiency and optical/ microwave experiments.

Maiman obtained his first laser action in ruby [7], [8] on May 15, 1960. The first laser, shown in Fig. 1, was a simple device. It consisted of a pink ruby cylinder, 1 cm diameter by 2 cm long. Both ends were ground and polished flat and parallel and were coated with evaporated silver. A 1 mm diameter hole in one of the silver coatings was provided to couple out the radiation. The ruby was mounted on the axis of a GE FT-506 xenon flashlamp, which was placed inside a polished aluminum cylinder (not shown in Fig. 1). As reported in [7] and [8], the first laser demonstrated both spectral and beam narrowing, together with lifetime reduction. In mid-July, three new ruby crystals, of much-improved optical quality, were received. When tested in the original setup, they provided more dramatic spectral and beam narrowing, a sharp threshold for laser action, and the relaxation oscillations characteristic of the ruby laser. A full report of Maiman's analysis and the exhaustive experiments conducted during the ensuing months is given in [5] and [6].

Unfortunately, the orderly publication of the first scientific report describing the first successful laser experiment was delayed by a comedy of errors. The principal error was the rejection of Maiman's article submitted to *Physical Review Letters,* pursuant to a new policy [9] holding that the maser field was so mature that maser articles no longer merited the speedy publication provided by *Physical Review Letters.* (Maiman had entitled his submission "Optical Maser Action in Ruby.") However, the work was published in two British Journals, and a well-attended press conference was held on July 7, 1960.

References

1. I. Wieder, "Optical detection of paramagnetic resonance saturation in ruby," *Phys. Rev. Lett.,* vol. 3, pp. 469-472, Nov. 15, 1959.

2. A. L. Schawlow and C. H. Townes, "Infrared and optical masers," *Phys. Rev.,* vol. 112, pp. 1940-1949, 1958.

3. A. L. Schawlow, "Infrared and optical masers," in *Quantum Electronics,* C. H. Townes, Ed. New York: Columbia Univeristy Press, 1960, p. 557.

4. T. H. Maiman, "Optical and microwave-optical experiments in ruby," *Phys. Rev. Lett.,* vol. 4, pp. 564-566, June 1, 1960.

5. T. H. Maiman, R. H. Hoskins, I. J. D'Haenens, C. K. Asawa, and V. Evtuhov, "Stimulated emission in fluorescent solids II. Spectroscopy and stimulated emission in ruby," *Phys. Rev.,* vol. 123, pp. 1151-1157, Aug. 15, 1961.

6. T. H. Maiman, "Stimulated optical emission in fluoorescent solids I. Theoretical considerations," *Phys. Rev.,* vol. 123, pp. 1145-1150, Aug. 15, 1961.

7. —, "Optical maser action in ruby," *Brit. Commun. Electron.,* vol. 7, pp. 674-675, Sept. 1960.

8. —, "Stimulated optical radiation in ruby," *Nature,* vol. 187, pp. 493-494, Aug. 6, 1960.

9. S. A. Goudsmit, *Phys. Rev. Lett.,* vol. 3, p. 125, 1959.

Charles K. Kao

Fiber Optics and Telecommunications

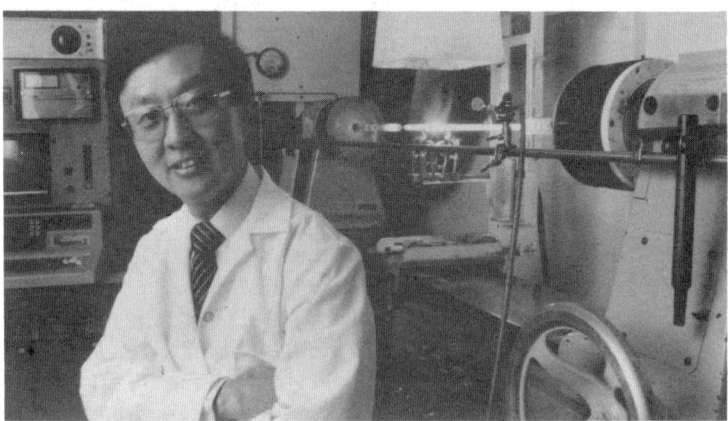

Charles K. Kao is recognized as the "Father of Fiber Optics" for his pioneering work at ITT's Standard Telecommunications Laboratories in England. In 1966, he was the principal author of a paper which accurately predicted the performance levels fiber optics could attain and prescribed the basic design and means to make it a practical and significant communication/transmission medium. His more than 30 patents to date reflect his substantial contributions to the technology required to realize his visionary concepts.

He obtained the PhD from the University of London in 1965, and has received numerous awards, including the Faraday Medal of the IEE (UK) in 1989, the International Prize for New Materials (1989) from the American Physical Society, the Alexander Graham Bell Medal of the IEEE (1985), and many others. He is author of Optical Fiber Technology II (IEEE Press) and Optical Fiber Systems: Technology, Design, and Application (McGraw-Hill)

Dr. Kao is Vice Chancellor and President of The Chinese University of Hong Kong. From 1982-1987 he was Executive Scientist/Corporate Director of Research at ITT's Advanced Technology Center in Shelton, Connecticut.

He was interviewed by Frederick Su.

You were instrumental in predicting the capabilities of fiber optics. Tell us about the history of that.

Briefly in the early 60s everybody was interested in looking for ways and means to use the optical spectrum for communication purposes. I was among the early people who were looking at the problem. About 1959, lasers were invented and for the first time we had the equivalent of a radio wave carrier at optical frequencies. And so the optical spectrum became available as a spectrum for communications. Optics is very attractive because its carrier wave is about 10,000 times higher in frequency than microwave. So, theoretically, one can get much more information onto it. So as a communication person, I was excited about that.

When we get into optical communication, one immediately faces several problems. One is how to get the signal from one point to the other. A transmission medium like free space has a great deal of limitation because of changing propagation conditions due to rain, fog, snow, etc. We started looking at alternative guiding mediums and, since I was from the waveguide side, the dielectric waveguide caught my attention. The problem was that transparent materials were very lossy. Some people theorized that perhaps the intrinsic loss of materials could not be reduced. It was then that I started to examine the basis for lossy material at optical wavelengths and decided that perhaps there was room for improvement, and that maybe one needs to look into the mechanism for losses in much more detail.

The losses turn out to be due to two sources. One is due to contamination or impurity absorption. The other is due to the intrinsic scattering loss. We did quite a bit of calculations and measurements and concluded that by improving the purity of a material like silica, one can reduce the loss considerably. Depending on the impurity concerned, one needs to achieve a purity on the order of one part per million to a few parts per billion. For instance, with iron, one has to reduce it to less than one part per million. With copper one has to reduce it to a few parts per billion.

There were intrinsic scattering losses, but they were something we couldn't do much about. We had a marginal effect by controlling the way in which the glasses were made. Fortunately, these losses were very low. Theoretically, it looked to be below one dB/kilometer at one micrometer wavelength.

This was in 1966-68 when we came out with a series of papers detailing these concerns. The one that particularly caught the attention of people was the 1966 paper published in the IEE Proceedings (UK). I understand it was reviewed by the Director of Research in the British Post Office. After it was published, the British Post Office took up this topic and we started a joint working situation between ITT Central Research Laboratory in England, the Standard Telecommunications Laboratory (STL) where I was working, and the British Post Office. The Post Office Research Station at that time was at Dollis Hill. We started working with glass

Terabit Technology

Charles Kao

We are asking the question, is 10^{12} bits per second possible? If not, why not? And if it is possible, how would that technology then impact the way we would do system design or system functional partitioning? If we can work at 10^{12} bits per second, which is about 1000 times the rate we can manipulate information bits currently, we are essentially delimiting the bandwidth limitations. So all those techniques which we have been using that conserve bandwidth may not be appropriate when we have bandwidth to burn.

Currently, communications systems are mainly handling voice traffic, and now we are talking about adding data to it. These are usually fairly low bit rate signals. The voice signal is 64 kilobits per second, and the data is at an even lower speed, but maybe one of the standards we're going to set will be data transmission at 80 kilobits per second. If we are handling that sort of traffic, we see that the trunk routes—that's city to city routes—have to have capacity already on the order of 500 megabits per second to handle the many, many conversations that go along these corridors. However, we can change the scenario and say that we're not only going to use the telephone to talk but to see as well. Using videophones, we now find that the bandwidth requirement is going to increase about 1000 times, because a good, high-quality video channel requires around a 100 megabit per second bit rate to handle it. If that is the case, and if the usage is as popular as the normal telephone, then we would envisage that for the trunk communication, we would need to increase the bandwidth to 500 gigabits per second rather than megabits per second. We really need the bandwidth. At the same time, fiber optics is offering a unique opportunity to create a transmission network like the interstate highway system; it is really capable of handling lots of traffic. Currently, the reason we're not using it for very high traffic is partially because the services are not here yet; wideband terminal equipment technology is not really mature. Electronics can barely work into the gigabit per second region, and if we want hundreds of gigabits, then we have a problem. That's where terabit technology is likely to make a hit in due course.

If we push the electron devices hard enough, the conventional devices like transistors can work up to 100 gigabits. And if we look at the optoelectronic devices, like detectors and lasers, they have an upper-frequency limit that currently is around, say, 10 gigabits. In optical devices, you can now generate a very short optical pulse at 10 femtoseconds (10^{-14} s), or a little shorter than that, even. All optical phenomena can go at very, very high speed. However, all optical logic gates that people have been working on are basically very slow. You can turn it on rather fast, but you can't turn it off fast enough. Its current speed limit is around 10^{10} to 10^{12} bits per second. We are on the border of being able to realize that sort of speed. We can't say that those demonstrations are especially meaningful because some of these things are done at the expense of excessive power. So if a switch can be made to operate at faster than one picosecond, but you have to supply a joule or two of energy to make it go, that won't be very practical in terms of a switching system. The field is wide open, and there are some promises. It's a very good research field because it touches the frontier of the technology.

Excerpted from *Optical Engineering Reports*, January 1987.

people—Pilkinton in England, Schott Glassworks in Germany, St. Gobain in France, and Corning Glassworks in the U.S. The best loss then was a few hundred dB/km. My team at STL worked out practically all the details of dimensional tolerance, geometry, waveguide structure, and so on. A series of papers, including the 1966 paper, demonstrated conclusively that optical fiber waveguides could be achieved in practice. By 1970 Corning published the first results claiming they had achieved a 20 dB/km loss, which was estimated as the highest loss that one could tolerate for some practical applications. The best loss previously was a few hundred dB/km. And so work started to grow, and it grew rather rapidly after 1974 when ITT leaped into the system applications of optical fiber communication systems. I returned to ITT and joined their U.S. effort, which was at Roanoke, Virginia.

At that time, fiber optics was regarded as a replacement product for copper wire in telecommunications if the fiber cost—if all the system component costs—could be lowered. However, even then, fiber's special properties, such as light weight, flexibility, no electromagnetic radiation, no electromagnetic interference, etc. ... well, all these properties were attractive for military applications. So initially there was quite a lot of funding through the Defense Department for doing early development work. That really helped— this was the mid 1970s—to consolidate R&D efforts until the first commercial production of fibers for commercial applications started to appear in the late 1970s.

Since the last time we talked you were at ITT Advanced Research Lab in Shelton, Connecticut. You were talking about terabit technology then. Did ITT just give up on terabit technology? Is it still a feasible project today?

I think that the terabit technology project, in my own analysis, was quite successful in spite of the fact that we had all this trauma of being sold to a French company, Alcatel.

I set up this project for several relatively interesting reasons. One of them was that I wanted to know whether there was a way of designing research projects that would turn out to be a management tool for managing R&D. The idea was that if we know something about a technology frontier, where it is and how fast it is moving, then as far as the R&D management is concerned, they can look at this and say, ok, we have determined the trend line and we can plot the various projects that we might be conducting elsewhere against that trend line. If it is too far from the trend line, then maybe we are not sufficiently ambitious, and therefore the project may be past the obsolescence point. Whereas if we push

There's a new word called globalization these days. In order to do scientific work and in order to do business these days, one needs a global network, and one needs to do R&D in a global fashion. Hong Kong happens to serve the world in a very significant way by being a key link between East and West.

it too close to the trend line, then we may be spending the effort too early because the products that can be derived from that type of technology may be too expensive. So it serves as a measuring yardstick for research management. And the terabit project had that usage.

But is there any other company pursuing that terabit goal?

Yes. Currently many people have designed projects that have the name terabit in them. What we did was look at the switching speed limits of various components to see how close we can work to terabit per second. And that helps to define and push, for instance, the bandwidth of an optical system into the low gigabit rate without too much difficulty. That's why you can see 1.7 Gbit systems already on the market. The other is to know that as we push up the frequencies for modulation, the techniques that one needs to use for modulation may have to be very different than what we are doing now. We need new device concepts. And that has led various groups of people to look at new devices, optoelectronic devices, that work on different principles than transistors and integrated circuits. We also know that as we push up in frequency, the intrinsic physics involved in the device must be carefully recognized as not being fully understood in the classical sense. One encounters problems that have to be analyzed more carefully using a quantum mechanical approach. So all together the project has served its purpose, not only in terms of coining the phrase "terabit," but also in terms of the intrinsic scientific value in defining what the field of terabit operation consists of.

ITT's Terabit Project wound down because ITT's operations were bought out by a French company?

Not wound down as such. Because the project was never intended toward, say, production of terabit devices. It was really a tool which had explicit purposes. One of them was to generate interest in terabit technology, and toward that end, there are many projects that are currently ongoing around the world that could be said to be stimulated by our project. People are looking at techniques of generating very, very fast bits and to see whether these very fast pulses can be handled—an example is soliton propagation. The project was not a means to an end. It was just a clarification of how we should deal with a very high bit rate environment.

And you think that will be forthcoming in the near future?

I don't know whether we want to go to those ultrahigh speeds. What we want to define is, what are the limitations as we go up in

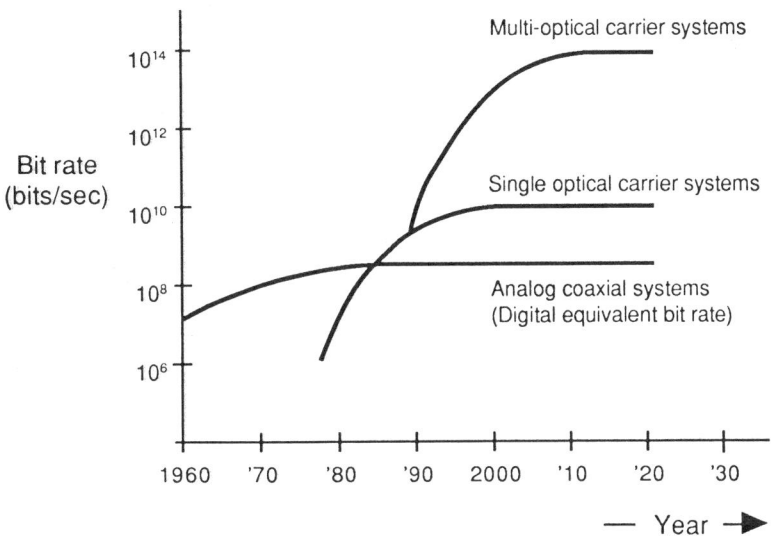

Multi-optical carrier systems

10^{14}

10^{12}

Bit rate
(bits/sec)

10^{10}

Single optical carrier systems

10^8

Analog coaxial systems
(Digital equivalent bit rate)

10^6

1960   '70   '80   '90   2000   '10   '20   '30

— Year →

*System bandwidth
projection. Illustration
courtesy of Charles Kao.*

sequential signal processing bit rates? And how high a bit rate one should aim for in practice? One should perhaps work in the 10s of gigabit range. There is also some question as to when parallel processing should be used, or how parallel and serial processing should be used simultaneously.

Did you hear about Alan Huang's optical computer at Bell Labs? What did you think about that?

That is a very good demonstration experiment. That if we try hard enough, one can get an optical computer to emulate a digital computer. I think it is a learning exercise just as the terabit project was a learning xercise.

Why did you decide to go back to Hong Kong?

There are several good reasons. One is that I have accumulated a vast amount of experience, particularly in networking. There's a new word called globalization these days. In order to do scientific work and in order to do business these days, one needs a global network, and one needs to do R&D in a global fashion. Hong Kong happens to serve the world in a very significant way by being a key link between East and West. My background could be utilized effectively to improve the service environment of Hong Kong. When this opportunity came up to help run a university, I took the chance. It's turning out very well.

I think Hong Kong is now involved in several key technologies and developments, one of which is biotechnology. It is a global technology-based business in which Hong Kong hopefully will play an important role.

And there is telecommunications. Hong Kong has the largest concentration of telecommunications circuits in the world. It has a few thousand international circuits connected to many different countries via satellite and submarine cables. There are three new transoceanic submarine cables coming to Hong Kong. The HJK or Hong Kong, Japan, Korea link; the Hong Kong-Taiwan link; and the Hong Kong-Singapore link. All of these are brand new fiber optic links that will go in, I think, by the end of this year. It turns out that Hong Kong is making a very significant telecommunications investment. The latest investment is for a fiber optic based cable TV system. It has a very interesting twist. The system is to carry telecommunications traffic at the same time as cable TV traffic. And that means this system is truly a broadband system of the BISDN (broadband integrated services digital network) type. With Hong Kong's existing network, which is highly fiber optic based, we have two overlapping networks, one an overlay system and one an existing system. Both of them can carry a variety of traffic.

So Hong Kong all of a sudden becomes perhaps the world's best network service testing laboratory. Nowhere else can we find such a density where we can do these service oriented experiments both for technical feasibility and pattern-use characteristics. I'm actually looking forward, technically, to contributing toward the design and development of future broadband systems.

What happens in 1997 when China takes over?

I have no idea. One has to be optimistic. People are doing all sorts of things to make Hong Kong viable. From my 2 1/2 years there, I see Hong Kong as a very important international service entity that really promotes trade and economic growth for the world in a very significant way, and hopefully, will continue to do so. Basically, the understanding is that Hong Kong should play the same role for the next 50 years after 1997. Even now, a very significant amount of investment money is scheduled, about 20 billion U.S. dollars, to be used in infrastructure development before 1997.

Willard S. Boyle and George Elwood Smith

The Inception of Charge-Coupled Devices

Willard S. Boyle

Willard S. Boyle retired from Bell Telephone Laboratories in 1979 and lives in Wallace, Nova Scotia. He is a member of the Research Council of the Canadian Institute of Advanced Research, and has been part of several advisory groups to government and universities. Prior to retirement, he was Executive Director of the Communication Sciences Research Division at Bell Labs. He holds 16 patents, including the charge-coupled device and the semiconductor laser. He is a Fellow of the American Physical Society, the IEEE, and the Canadian Academy of Engineering.

George E. Smith received the PhD from the University of Chicago in 1959. He was a staff member at Bell Labs from 1959 to 1969. He is a Fellow of the IEEE and the American Physical Society.

Both Boyle and Smith received the Franklin Institute's Ballantine Medal (1973) and the IEEE Liebmann Award (1974)

It would seem that the authors of the papers in this special issue have a very heavy responsibility. Much has been written about the mental processes that lead to innovation but this is one of the rare times that a group of people who themselves have participated in the act of innovation have been asked to shed some light on the factors which contributed to the generation of a new concept. It is not difficult to document the apparently important features; they have been told many times before. Indeed the telling and retelling with embellishments of "the day that etc." gradually takes on a ring of authenticity that eventually becomes fact in itself. On this occasion, however, we shall try to analyze once again just exactly what did transpire, with the full understanding that with all the best effort at objectivity, it may still not be an accurate account of the subtle interplay between people that certainly played a most important role in our own work.

The charge-coupled device concept is one that is basically a structure that called upon existing technology and was stimulated by the analogous work that preceded it in magnetic bubbles. All the ingredients necessary for the innovation were found within Bell Laboratories. The now well-known work at the Philips Eindhoven

Laboratory by Sangster[1] and his colleagues on the Bucket Brigade was unknown to us. The importance of the Bucket Brigade as a member of the family of charge transfer devices is well recognized, and we want at this point to recognize the important contribution to device technology that has been made by the work at Philips.

In late 1969, the device area of Bell Laboratories was strongly oriented towards innovation and exploratory development of new devices. Strong efforts were being directed towards such fields as IMPATT diodes, GaAs lasers, nonlinear optics, solid-state lasers, holography, Gunn diodes, magnetic bubbles, and the silicon diode array camera tube. Of particular interest was the work on magnetic bubbles. It was apparent that a radically new approach to signal processing was being brought into existence.

Those of us who were working on semiconductor integrated circuits looked at the work in magnetic bubbles with some awe. In the magnetic bubbles it seemed that a new class of devices was in the offing in which there was a very natural and appealing way of storing and manipulating bits with a one-to-one correspondence to the digital format. It could be argued that semiconductor integrated circuits, important as they were, had not yet broken away from the circuit concepts that had evolved through the use of discrete components. The late Jack Morton, who was at the time Vice-President of the Electronics Technology area, was a strong proponent of the magnetic bubbles program but felt, at the same time, that surely there must be some analogous devices using semiconductors. He was both persuasive and Vice-President, so his persuasions to develop a semiconductor bubble-type device

A certain amount of arrogance is essential in carrying forward an idea. It is surprising now in retrospect the number of people who either were quite negative and had reasons to suggest it would not function as described, or claimed that it would be of little interest and no better than some already existing device.

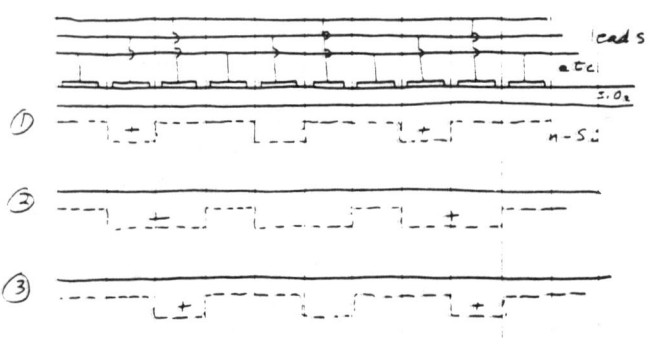

Firure 1. Reproduction of the notebook sketch of the first three-phase charge-coupled device. ©1976 IEEE.

Current and Future Impacts of the CCD
Willard S. Boyle

What is the great advantage of your invention?

The great advantage is in the video area. It has had a tremendous impact on optical astronomy, in the sense that the video devices that you can make with the CCD are about 10 times more sensitive than the similar kind of thing you can get with photographic plates. It has extended the range of astronomical telescopes in sensitivity by about a factor of 10, or 3 times in distance. Not only that, but since the signal is coming out in a form which you can manipulate directly with processing, you can do things like remove the background in the sky that you are looking at, and you can integrate for a long period of time. Using these principles, astronomers have made some quite major discoveries in the last five years or so. This sensitivity and processing capability make it very useful on some of the satellite imaging devices and some of the deep space probes. Also it is at the heart of the new Hubble telescope. There are arrays of these CCD sensors in the telescope that will allow people to sit back here on earth and watch what is happening in deep space.

And then there is the impact on video cameras.

Yes. The sensing element in a video camera is a CCD. If you look at the modern camcorder you will see that many of them have a little CCD stuck on the side of it. The main image sensor is a CCD imaging device. The advantage is that the camcorder can be made very small. If you tried to do the same thing with a typical vidicon, it would be a much bulkier device and require much more power. Also, since the CCD is a semiconductor device, it is not subject to burn-in, which is required for vidicon cameras. If you inadvertently point the camcorder at a flash bulb going off, it wouldn't damage the CCD in the same way that it would damage the vidicon. To sum up: the CCD is small, it is low power, it has very good light sensitivity, and it is rugged.

What do you see in the future for the CCD?

I think the CCD's main applications in the future will be using it as a camera for all the reasons I have mentioned above. But predicting the future is always difficult. At the time we invented the device we had no idea that imaging was going to become the dominant application. Its applications will continue to be dominated by its uses as an imaging device in many, many forms. An example would be imaging in robotics. It's extremely important to be able to have a signal that you can easily manipulate and that can control your robot arm. Other areas include spectroscopy, to be able to have nice, compact spectroscopes; in microscopy, to be able to identify and sort for medical purposes, to be able to do blood counts and things like that digitally. There is just tremendous opportunity to manipulate visual images. It is a very rapidly growing field.

Interviewed by Frederick Su

received rather close attention. The encouragement from management was there. Indeed, it was more than encouragement at times and it seemed that unless we came up with something comparable, future funding of the exploratory programs in semiconductors might be in jeopardy. Certainly this was only a perceived threat, but it does convey the atmosphere at the time.

Another very important ingredient was the development program for the silicon diode array camera tube (2). One of us (G. E. Smith) had been deeply concerned with the material and device technology required to fabricate defect-free diode arrays in silicon. These diode arrays had to have both the light sensitivity and the charge storage capability to provide an imaging device of acceptable quality for the PICTUREPHONE imaging tube. The charge storage on individual diodes had been particularly troublesome. However, success was at hand, and it was possible to routinely fabricate hundreds of thousands of diodes on a single chip without a single defective diode. Moreover, the charge could be stored in one of these diodes for periods approaching a hundred seconds.

It was in this atmosphere that the charge-coupled device was born. During an afternoon discussion between the authors lasting not more than an hour, the structure and some preliminary ideas concerning applications of the device were developed. The train of thought evolved as follows.

First, the semiconductor analogy of the magnetic bubble is needed. An electrical dual is a packet of charge. The next problem is how to store this charge in a confined region. The method which came to mind was the metal-oxide-semiconductor (MOS) capacitor in depletion. This forms a potential well at the surface into which one can introduce charge (or not) to represent information in the same way that the presence (or absence) of magnetic bubbles on a site formed by a permalloy pattern represents information. The last problem was to find a way to shift the charge from one site to the next, thereby allowing manipulation of the information. This is solved by placing the MOS capacitors very close together in order to easily pass the charge from one to the next by applying a more attractive voltage to the receiver. This completed the basic invention, and the attendant frills and possible applications came rapidly after. The first notebook drawing of the device, shown in Fig. 1, depicts the basic 3-phase configuration and is still valid in describing today's device.

It seemed almost too easy and straightforward, so, having had past experience in seemingly brilliant ideas which subsequently fizzle, we allowed the idea to remain just that for a few weeks. In talking with colleagues the reaction was varied, ranging from "I should have thought of that" to lengthy lists of reasons why it would not work.

At this point we want to observe that a certain amount of arrogance is essential in carrying forward an idea. In talking about the device with others, it is surprising now in retrospect the number of people who either were quite negative and had reasons to suggest it would not function as described or claimed that it would be of little interest and no better than some already existing device. Although each of us had more than our share of ideas that were of no consequence, we had also experienced the frustration of having dropped a proposal only to have it independently brought forward successfully by someone else. Our frame of mind at this time was such that we had confidence that our idea was sound and important regardless of how negative a few of the comments from our colleagues might be.

Finally, it was decided to go ahead and fabricate a device to show experimental feasibility. In less than a week, masks were made and devices were fabricated, mounted, and tested. The photo in Fig. 2 shows the first device which consisted of a simple array of 0.1 mm × 0.1 mm MOS plates placed in a row with three-micron spacings between them. Charge was injected by avalanching the first plate and, after a series of transfers, the charge was detected by measuring the current produced by injection into the substrate. This activity took place in October 1969.

Figure 2: First experimental array of field plates used to demonstrate the principles of charge transfer. © 1976 IEEE

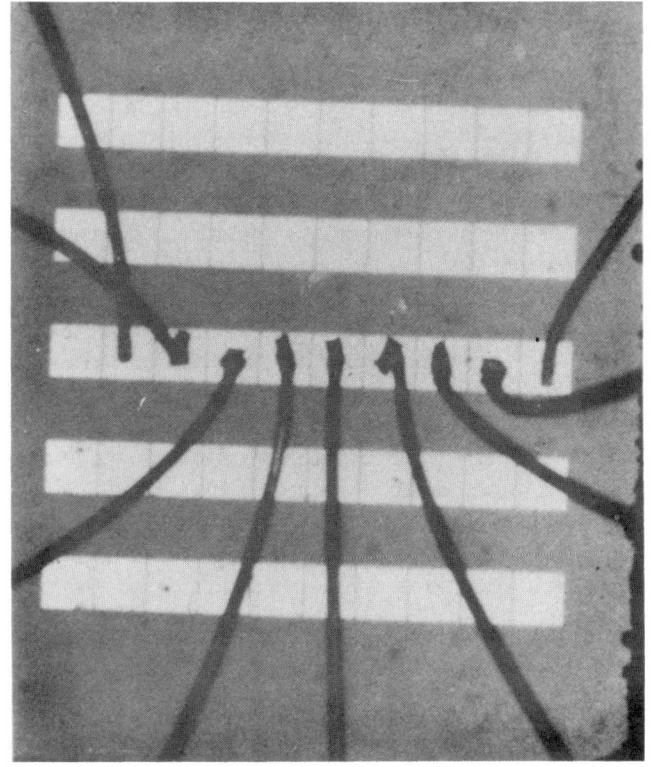

A beehive of activity and intellectual excitement took place immediately after this. The idea was infectious and many others made substantial contributions in the weeks to follow. Dawon Kahng developed the idea of a built-in barrier under one side of a plate to make a two-phase device (3). A three-phase device can be used for a linear device or an array in which the charge goes in one direction but, for a serpentine memory array, a two-phase device is needed to avoid crossovers.

Eugene I. Gordon conceived of a display device in which the video charge pattern is read in serially and then injected into the substrate in a parallel manner to produce a display via radiative recombination. Harry J. Boll and C. Neil Berglund reinvented the Bucket Brigade, being unaware of F. L. J. Sangster's prior work and also recognized the importance of a fat zero. Boyle and Smith, recognizing that complete charge transfer can be limited by surface-state trapping, circumvented the problem with the buried-channel concept. This device uses a structure in which the charge is not stored at the semiconductor-insulator interface but in the bulk of silicon. This can be done with a special doping profile and then the complete transfer is limited by the much less numerous bulk states. Many others contributed with ideas too numerous to mention.

The first public announcement was made at the New York IEEE Convention held in March 1970. Boyle was a member of a panel discussion on the future of integrated circuits and came equipped with one vuegraph showing the basic operation of CCDs. He spent less than five minutes explaining the device but many people understood immediately, as evidenced by the lively discussion which followed. It was also picked up by the press and covered in several of the trade publications.

The ideas were subsequently reported by Boyle and Smith (4) in the April 1970 issue of the Bell System Technical Journal in which charge storage in an MOS capacitor was discussed, the basic three-phase structure described, and possible applications outlined. Storage times and the three factors determining transfer efficiency were discussed; namely, the limitations imposed by diffusion, and the field enhancement resulting from the change in surface potential with charge density. Overcoming the limitations by using geometrically induced fringing fields was also mentioned. Several applications were discussed in the papers. One was to use the device as a serial shift register or memory with p-n junction input and outputs. The background of bubbles naturally led to thinking of the shifting of digital information in the form of charge or no charge, but the diode array influence soon made the analog application of imaging come to mind, both linear and area. Here a parallel input is produced by light making electron-hole pairs and then the analog information read out in a serial fashion. Use of the device as an electrical analog delay line was envisioned but the full significance of this was not realized at the time and perhaps still is not. Two-dimensional arrays and logic were also mentioned.

The first experimental results were reported in a companion paper by Amelio, Tompsett, and Smith (5) in which storage times

Figure 3: Image produced by an 8-bit linear array of CCD elements. Scanning from left to right was obtained mechanically. ©1976 IEEE.

of 16 s and transfer efficiencies of 98 percent were reported.

The first regular technical paper was presented by Smith at the Device Research Conference held in Seattle that June. A paper by Tompsett, Amelio, and Smith (6) on the first device with diode inputs and outputs and integrated interconnects came out in August. This was an eight-bit device which had a transfer efficiency of 99.9 percent at 150 kHz. The device was also used to demonstrate imaging for the first time and even though its eight-element accuracy left something to be desired (see Fig. 3), the principle was established. By this time, many other workers had started work on CCDs and activity was accelerating at a high rate. A high level of activity has remained and, at this time, appears that the device will find a permanent place in electronic systems. Area imaging devices with the full commercial format have been demonstrated, and several area and linear devices are available commercially as are digital memory devices containing up to 16 kbits. Perhaps the most innovative applications are yet to come.

References

(1) F.L.J. Sangster, "The bucket brigade delay line, a shift register for analog signals," Philips Tech. Rev., vol. 31, pp. 97-110, 1970.

(2) M.H. Crowell and E.F. Labuda, "The silicon diode array camera tube," Bell Syst. Tech. J., vol. 48, pp. 1481-1528. May-June 1969.

(3) D. Kahng and E.H. Nicollian, "Monolithic semiconductor apparatus adapted for sequential charge transfer," U.S. Patent No. 3,651,349.

(4) W.S. Boyle and G.E. Smith, "Charge coupled semiconductor devices," Bell Syst. Tech. J., vol. 49, p. 587, Apr. 1970.

(5) G.F. Amelio, M.F.T. Tompsett, and G. E. Smith, "Experimental verification of the charge coupled device concept," Bell Syst. Tech. J., vol. 49, p. 593, Apr. 1970.

(6) M.F. Tompsett, G.F. Amelio, and G.E. Smith, "Charge-coupled 8-bit shift register," Appl. Phys. Lett., vol. 17, p. 111, Aug. 1970.

How does the CCD work?

The CCD works by the generation of an electron-hole pair via the photoelectric effect (Figure 1).

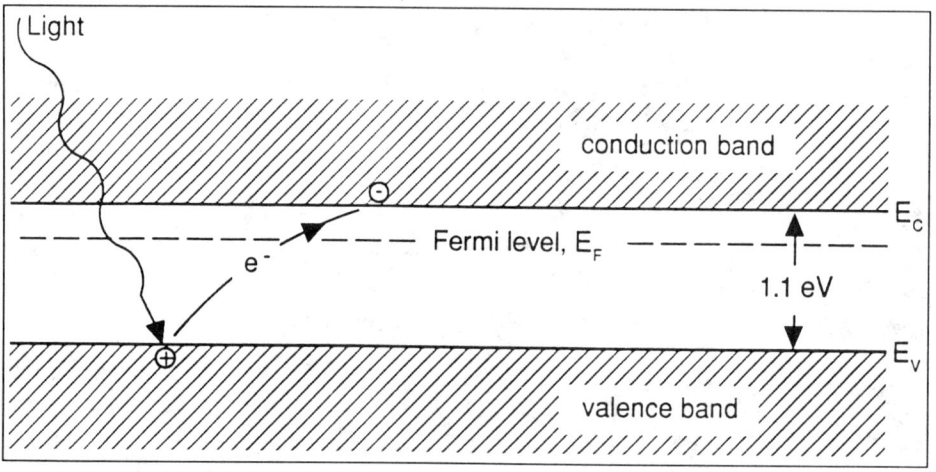

Figure 1. The photoelectric effect in silicon.

The holes are trapped in potential wells in the CCD (Figure 2a). By varying the voltage at the electrodes, the charge can be transferred along the surface by moving the potential wells (Figure 2b and Figure 3).

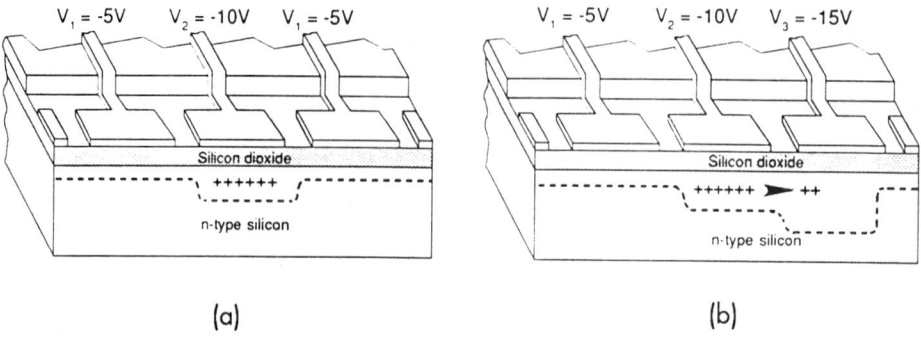

Figure 2. Cutaway view of a CCD in (a) the storage condition and (b) the transfer condition. From "Charge-Coupled Devices—A New Approach to MIS Device Structures," by Boyle and Smith, IEEE Spectrum, July 1971. © 1971 IEEE.

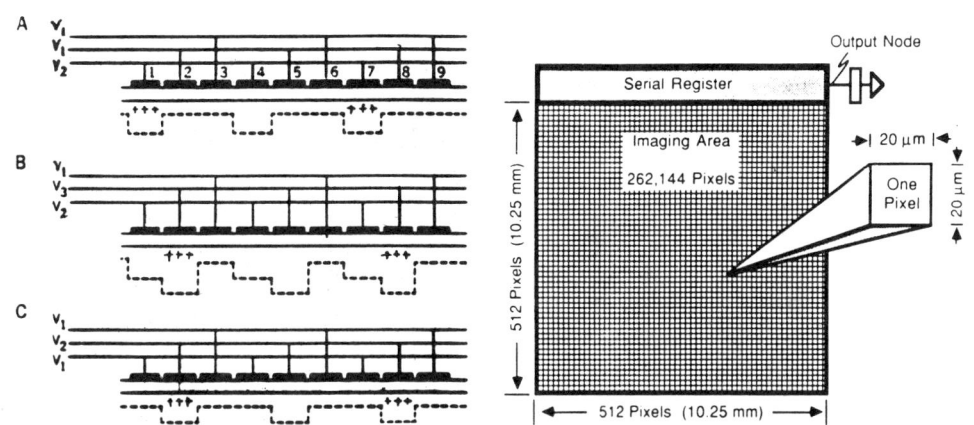

Figure 3. The shifting of the potential wells in a 3-phase CCD. From "Charge-Coupled Devices— A New Approach to MIS Device Structures," by Boyle and Smith, IEEE Spectrum, July 1971. © 1971 IEEE.

Figure 4. Typical 512 ×512 CCD. Reprinted with permission from Charge-Coupled Devices for Quantitative Electronic Imaging, Photometrics Ltd., 3440 East Brittania Drive, Tucson, AZ 85706.

The concept can be extended to a two-dimensional array, typically 512 × 512 pixels, or 512 × 512 potential wells. The two-dimensional array is called the parallel register (Figure 4).

The light that forms an image falls on the parallel register. Therefore the array of potential wells is an image formed not of photons but of electrical charge instead. Each row of the parallel register—with each element (we can also call it a pixel or potential well) storing charge in proportion to the amount of light that hits it—is then shifted upwards one at a time into the serial register. Each element of the serial register is then shifted to the right one at a time to the output node. This is shown in Figure 5. Note that in going from Figure 5b to 5c we have neglected to show the clearing of the serial register.

Typically, the charge from the output node is fed into an electron gun. The electron gun scans a phosphor screen as in a TV to form the image.

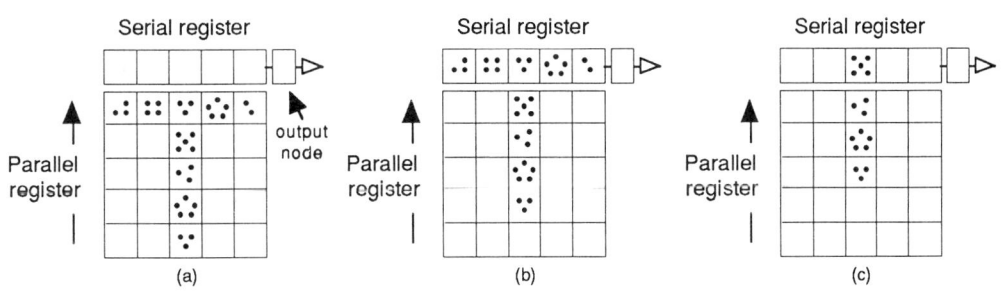

Figure 5. The parallel and serial shift registers. Video cameras use a more complicated shifting arrangement.

Allan Cormack

The Development of Computerized Axial Tomography

Allan Cormack, Tufts University, was co-winner of a Nobel prize in 1979, with Godfrey Hounsfield, for the invention of Computerized Axial Tomography. He was interviewed by Frederick Su.

When did you first start to look into the field of medical physics? To envision the CAT scan?

That was in 1956.

How did that come about?

It is a long story. In those days, at the Groote Schuur hospital in Capetown, the one that Chris Barnaard made famous with his heart transplants, they had only one medical physicist. He quit and went to Canada. I think I was the only person in South Africa at the time who knew much about nuclear physics and how to handle radioactive isotopes, service film badges, and such things. So I was asked to spend a day-and-a-half a week looking after the isotope program and related things at the hospital. I was put in the radiology department of the hospital. So I was there to see how they planned treatments for x-ray therapy. They did this by superimposing isotope charts representing beams coming into the body from different directions. They then looked at the resulting distribution and adjusted it to suit the radiologist. But the thing that really bugged me was that these isotope charts were calculated on the basis of an assumed uniform absorption coefficient, and clearly lung and bone have different absorption coefficients. How could they possibly do decent therapy planning without being able to take into account the variations in the absorption coefficient throughout the body? So my first goal was to obtain a map of absorption coefficients for this purpose. But then it occurred to me that the map itself might be interesting, but I didn't realize at the time just how very interesting it would be.

So you knew that there were different absorption coefficients for different parts of the body, the soft tissues and the hard tissues. Where did you go then?

The question was, how do you find this out? You have to do it by making x-ray absorption measurements exterior to the body.

Are these on live patients or on cadavers?

> *The thing that really bugged me was that these isotope charts were calculated on the basis of an assumed uniform absorption coefficient, and clearly lung and bone have different absorption coefficients. How could they possibly do decent therapy planning without being able to take into account the variations in the absorption coefficient throughout the body?*

No, this was just all theoretical in the beginning. First, I had to develop a mathematical way of doing this. The problem itself is known as Radon's problem, which I had never heard of. Namely, how do you reconstruct a function knowing its average values along straight lines in the plane.

So, is that the basis of the CAT scan?

Yes. You start by taking a series of absorption measurements through a cross-sectional slice of the body. Fire some fine beams along lines which intersect that cross section (Figure 1) and measure the absorption along those lines. That gives you the average values of the absorption coefficient along those lines. The question then is to go from the average value to the point-by-point variations throughout the body, i.e., throughout that particular slice.

That means rotating the source about the body?

Right.

How many degrees do you rotate it for each picture?

In my first experiment?

Sure.

In the first experiment I fired a series of parallel rays every 7.5 degrees. I used a sample that was symmetrical about an axis and so I only had to do 180 degrees. The sample was a lucite disk with an aluminum ring around it and, inserted in the disk, were two smaller aluminum disks. The outer aluminum ring was supposed to represent the scalp and the lucite was supposed to represent normal tissue. The two inner aluminum disks were supposed to represent two tumors. I chose this configuration because of a case that had come to my attention. Someone had just been treated for a malignant tumor in the head, successfully it seemed. But the patient succumbed because there was a second smaller tumor that had escaped detection. Looking back, the radiologist said, "Oh yes, there it is but we didn't see it." It seems ridiculous these days, but the idea in that first experiment then was to see a smaller tumor if there was a bigger one that was somewhat obscuring it.

So these disks and the lucite had pretty much the same values for their absorption coefficients as tissue and bone?

No, it is more complicated than that. I had been looking at the radiological literature before this and I didn't see where this somewhat mathematical stuff would find a place. I also knew that people were doing what is now called PET (positron emission

I did not realize at the time how hard it was, and still is, to see a tumor with an ordinary x-ray. Because you see, the different thicknesses of bone in the skull complicate matters tremendously. It was incredibly difficult to see anything. And, suddenly, things that had been hard to see were now clearly visible to radiotherapists.

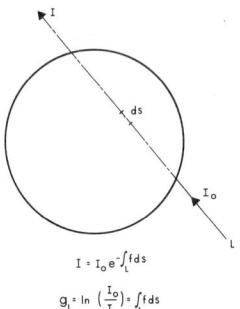

$$I = I_o e^{-\int_L f \, ds}$$

$$g_L = \ln\left(\frac{I_o}{I}\right) = \int_L f \, ds$$

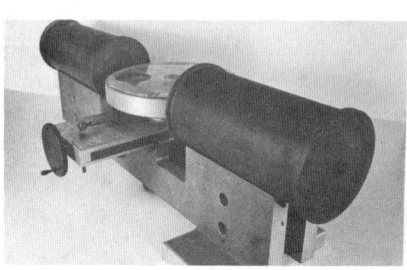

Figure 2. Prototype scanner of 1963. Cost: ~$100.

Figure 1. If L represents a fine beam of x-rays traversing a body, a measurement of the incident intensity (I_o) and the emerging intensity (I) yields the absorption along the line L. From measurements made on many lines intersecting the body, one can deduce the variation of the absorption coefficient from point to point within the body.

tomography) scanning in a very crude way. The problem I was working on would provide the solution to the PET scanning problem, which had not been found at that time. So I chose the absorption coefficients such that the ratio of the absorption coefficients of the lucite and aluminum for 1 MeV ^{60}Co gamma rays would approximate the sort of conditions you might find in PET scanning. I was hoping the people who were working in PET scanning in those days, which really meant just taking two orthogonal views, might be interested in picking the subject up. In fact, I showed it to a couple of people who were working in PET scanning and they were not the least bit interested.

What year was that?

That would have been 1963.

In any kind of problem like this you formulate the problem and you attack the problem and you go down a lot of blind alleys. Now, somewhere along the line you must have had an idea of the correct way to attack the problem and make that solution a reality. Did that occur to you?

Well, the problem was easy enough to formulate as a mathematical problem and, as I said, it is a problem now called Radon's problem. I had not heard of it at that time. So I had to set about solving the problem *ab initio*. There were three somewhat different types of solutions, one of which was obviously unstable for computation. The two others were stable solutions.

I'd like to get some of your own perceptions on the impact the CAT scan has made?

It is enormous. People are seeing stuff that they were totally unable to see before. The people who do radiology to detect soft tumors of the head were the ones who were first attracted to CAT scanning. And it easily exceeded their expectations as a tool for diagnosis. Then, there was some guy at the Massachusetts General Hospital here who, shortly after CAT scanners came out, said, "Oh yes, it may be fine for the head, but it will never work for the chest, for example." And, of course, as you know, it is used to image all parts of the body. I did not realize at the time how hard it was, and

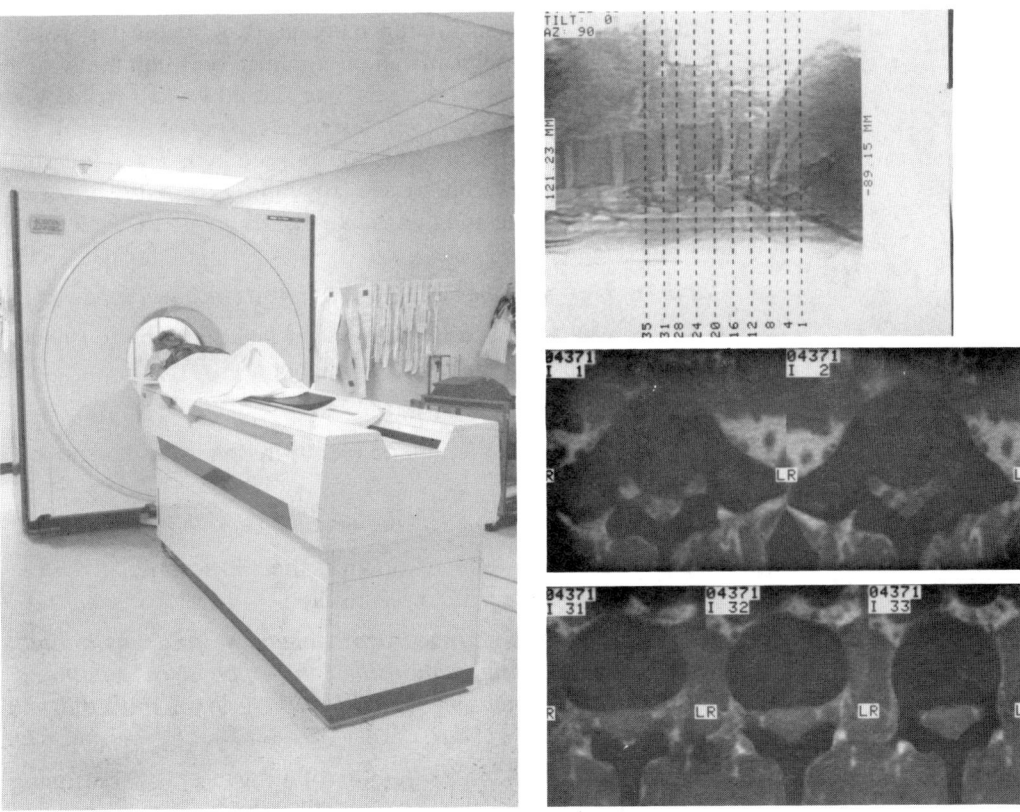

Figure 3, left: A GE 8800 CT/T (computerized tomography/total body) machine. (Photo: Gail Bertolini-Su).

Figure 4, top right: A sagittal view of a patient's back from the 1st sacral to the 3rd lumbar. There were 35 images taken. It takes almost 10 seconds for the x-ray tube to go around the patient. These were done in 5 mm thickness cuts spaced every 3 mm—so there is an overlap. It takes 35-40 seconds for the computer to reconstruct each image. Newer machines and software permit a much faster scan and reconstruction of the image. (Photo courtesy of St. Joseph Hospital, Bellingham, WA).

Figure 5, middle and bottom right: Cross-sectional or axial views of the vertebrae and intervertebrae soft tissues. This is a positive taken from a negative, so compared to a regular x-ray picture, the bone (or denser material) appears darker than the soft tissues. Numbers 1, 2, 31, 32, and 33 refer to the specific image taken and are referenced along the body by the sagittal view (Figure 4). The R and L refer to the right and left sides of the patient's body. (Photos courtesy of St. Joseph Hospital, Bellingham, WA)

still is, to see a tumor with an ordinary x-ray. Because you see, the different thicknesses of bone in the skull complicate matters tremendously. It was incredibly difficult to see anything. And, suddenly, things that had been hard to see were now clearly visible to radiotherapists. That provided the first impetus for buying CAT scanners. People like Dr. Giovanni DiChiro in the NIH saw the value of this process. It just saves an awful lot of diagnostic surgery.

How does a CAT scan distinguish a tumor from normal healthy tissue?

It is basically a small difference in density. A couple of percent, I believe. And the CAT scan can discriminate between the two. If you really want to appreciate this, you should go to a radiologist friend and get him to show you a standard x-ray of the head. Go ahead and see how much detail you can see in it. You'll find that it is enormously difficult to see anything except the grossest features such as bone and things like that. But with the CAT scan, you can (Figures 4,5).

Has the resolution of the CAT scan improved since you invented it?

Oh yes, all I made was essentially a feasibility study, I never attempted to make a commercial scanner.

What was Hounsfield's role in this?

He apparently did this independently in Britain. He had no role in my work at all.

Did he do the same thing as you, or did he develop the machine?

He produced a machine known as the EMI scanner; it was the first one that got the radiologists excited.

Your invention has changed world medicine and has brought about a closer working relationship between physicists, engineers, and those working in the medical field. Where do you think that will lead us in the years ahead?

Goodness knows. I like to think of it this way. CAT scanning depends really on the physics of the 1890s when x-rays were first discovered. PET (positron emission tomography) scanning was developed based on the physics of the 1930s when positron annihilation was discovered. And magnetic resonance imaging (MRI) is based on the physics of about 1950 when NMR was discovered. Who knows where we might be in the future? Only time will tell.

Shaoul Ezekiel

Optical Gyroscopes

Shaoul Ezekiel is a professor at the Massachusetts Institute of Technology, where he conducts research in optical frequency standards, high resolution laser spectroscopy, and optical gyroscopes. The interview was conducted by Roy Potter, SPIE Technical Consultant.

Why are you interested in gyroscopes?

Well, for two reasons. First, it is an incredibly sensitive measurement, a few parts in 10^{18} just for a regular old rotation sensor for navigation. If you want to do any better geophysics and relativity, you really are talking about parts in 10^{22} or 10^{23}—it's a fascinating area to get into. Second, one of the departments that I'm associated with is Aeronautics and Astronautics, and I'm always on the lookout for an area of interest to the department. Since the mechanical gyroscope and inertial navigation were developed by Draper at MIT, I thought it would be appropriate if some work also went on with optical gyroscopes at MIT. We got started on a passive resonator approach where we used the same ideas as the clocks; we locked lasers to resonances. We tried to do the same thing with passive ring cavity, where the clockwise resonance would be different from the counterclockwise resonance in the presence of rotation, except that here we are talking about shifts of millihertz instead of kilohertz and megahertz. So it was an interesting kind of spectroscopy. We called it ultrahigh resolution spectroscopy of resonators. The trick in that was to get rid of laser jitter by using one laser and two Bragg cells. If you had two lasers it would never work because the jitter in each laser would be too large. We can stabilize lasers, but to measure millihertz differences you really have to get the laser jitter to be very small (on the order of millihertz). By using Bragg cells you make the laser jitter common mode to both clockwise and counterclockwise measurements. But you have to drive the two Bragg cells with good oscillators.

Bragg cells are almost ubiquitous devices.

Oh yes. We use them a lot in the lab. We have one experiment where we use seven or eight Bragg cells. They're very, very useful.

Would you outline the purpose of the Bragg cell?

In an experiment with Bragg cells you can use one laser instead

Excerpted from _Optical Engineering Reports_, September 1984.

of two. You accomplish that by taking a laser beam, splitting it into two, and then sending each beam through a separate Bragg cell, which is driven by a good oscillator. This means you must use frequency synthesizer oscillators, and you get no relative jitter at all this way. The jitter is common mode and, in fact, we checked this before we started any of those experiments to see that, indeed, we didn't have any jitter introduced by the Bragg cells.

How much of a frequency shift do you get from the Bragg cell?

From the Bragg cell you get, for the one we normally use, 40 megahertz. In our iodine experiments, we use up to 200 megahertz. With the Raman clock we use 1.7 gigahertz, and you can get even higher than that. We use them mainly for frequency shifting. But the trouble is that the deflection angle depends on the frequency, so you have to be careful when you try to change the frequency too much because you get a change in angle.

Now we're back on the resonator.

Our interest is in making a big resonator for geophysics or relativity-type experiments.

What do you mean by "big"?

On the order of 100 meters on a side, something like that.

A resonator with dimensions of 100 meters or so?

Yes, using maybe a 10-watt argon laser as the source; then you can do interesting things. You can tell whether the earth is wobbling.

You mean you have an experiment in mind to test that?

Of course.

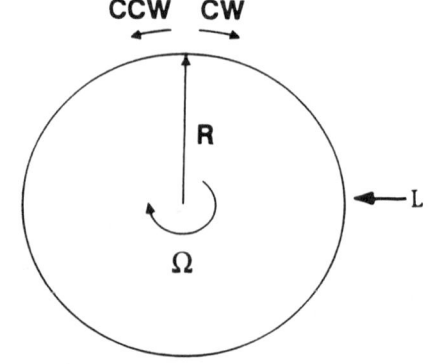

Optical rotation sensors are based on the Sagnac effect. An accurate solution requires a general relativistic formulation, but a simple expression can be found as follows: Over a period Δt,

the CW beam traverses $2\pi R + \Omega R \Delta t$,

the CCW beam traverses $2\pi R - \Omega R \Delta t$.

The path difference, D, is then

$$D = CW\ path - CCW\ path = 2\Omega R \Delta t$$
$$= 2\Omega RL / c,$$

where $\Delta t = L/c$ is the time it takes light to traverse the circumference L.

We know the earth is wobbling, but we want to know how much.

Whether the earth is speeding up or slowing down, the effect of tides on the earth's rotation.

How would you do such an experiment?

Very carefully!

You don't want to shake the earth any more!

No. The problem is to find a quiet place to do it. That's why the work couldn't be done at MIT; it's too noisy. Also, once you go outside MIT you can't generally use students. You have to have full-time people.

Do you have any prospective locations for your project?

Right now we are hoping to piggyback on the research that's going on at the sites that are being looked at by people searching for gravity waves because there they have to build an interferometer that's about 5 kilometers on a side. If we could somehow talk them into using a square instead of just an "L," then we could use the setup or a small part of that setup for the gyroscope. [See Kip Thorne interview.]

You could combine them. It would be one facility.

A ring laser gyro consists of a triangular path ring laser, which is fabricated in a quartz block for stability. There is no external light source as in a fiber optic gyro. Three mirrors define the light path, which is identical for the two directions, so the optical frequency is also identical, except for effects of rotation. A high voltage discharge in HeNe gas over part of the path provides the gain; and, after passing through one of the mirrors, samples of the two beams are combined on a pair of detectors. When the RLG is rotating, the frequency of the two laser beams is different, with the difference being proportional to rotation rate. Measurement of the detector output frequency then is a measure of the rate of rotation. (Illustration courtesy of Honeywell.)

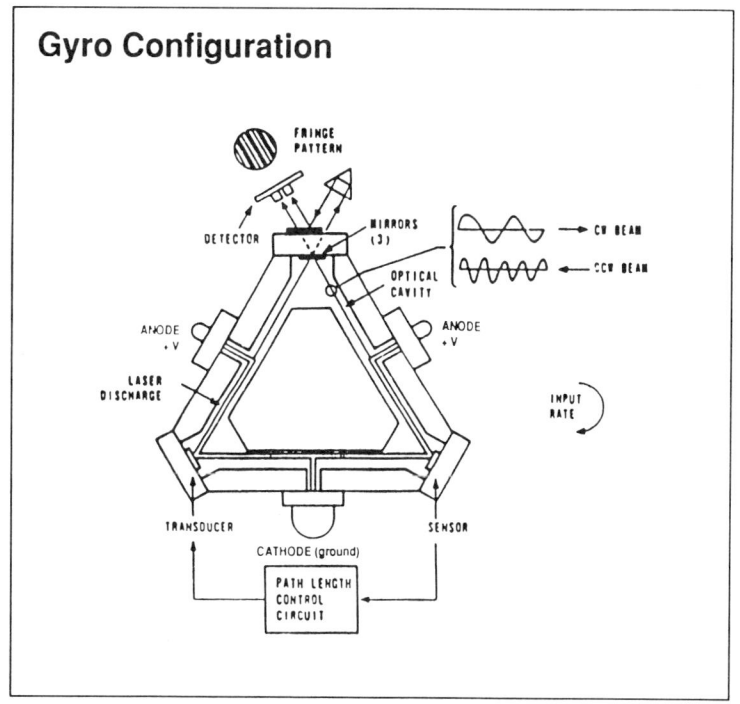

Gyro Configuration

That's right, because in all these facilities you need real estate with air conditioning and electronic systems. So it's easier to combine it with another project than to just do it by yourself.

Are there any prospects for such a joint facility?

Well, the NSF is sponsoring the gravity wave project. Maybe they would consider sponsoring the rotation project as well. But to do it right we still have to build a three-axis system. One axis is easy; then you have to build the other two axes. So you must find a mountain or something, and set it up on the side of the mountain. Nothing comes easy. When low-loss fibers became available, I became interested in a fiber interferometer gyroscope using the same ideas we used for the stabilization of lasers—closed-loop techniques using Bragg cells. So we proposed that for the fiber interferometer. Now we're also working on a fiber resonator, which is the same as a mirror resonator but using fiber instead of mirrors, again using Bragg cells and closed-loop techniques. But to get good performance a fiber resonator really needs very good fiber with respect to polarization. You have to have a good polarization-maintaining fiber; otherwise any small perturbation in the resonator would give you a change in birefringence, and that would give you nonreciprocal phase shifts. So what we're really developing is a very good temperature sensor!

You mean the temperature affects the polarization due to the birefringence?

Yes. In fact, we turned the fluorescent light on in the lab with the fiber resonator uncovered, and we saw an effect. And it wasn't stray light falling onto the detector. By just putting a little black cloth over the fiber, the effect would go away. The fiber was actually enclosed in a Lucite box. So the light was going through the Lucite, heating up the fiber to generate change in birefringence. When you're looking for effects on the order of one part of 10^{17}, almost anything affects your experiment.

Would you comment on optical gyros and resonators vis-à-vis mechanical gyros?

Ring laser gyros now are doing very well. Boeing is using the Honeywell gyros for their 757s and 767s. Litton gyros are being used by Airbus and are also doing very well. Ring laser gyros are here to stay and competing with mechanical gyros for inertial guidance. What's going to happen in the low-cost area, I don't know. Fiber gyroscopes are being pursued by many groups. Also, some companies are looking at fiber resonators or waveguide resonator gyroscopes. All kinds of exciting things are happening.

And somehow because of optical communication and the development of semiconductor lasers and specialized fibers, it's inevitable, at least the way I see it, that something is going to come out in fiber optic gyroscopes. Whether we can compete with ring laser gyroscopes is again to be seen because, in the meantime, ring laser gyroscopes will also become more refined and less costly. So it will be very exciting to watch the development of the new gyroscopes based on fibers.

Some years ago there was a suggestion that fiber optics actually could be made part of an active medium. Has anything come of that?

To make a fiber ring laser gyro? No, I think the scattering problem would be tremendous. It's bad enough with three or four mirrors in a ring laser. I think the passive approach is probably the way to go because you have a handle on the error sources. Often if you can see an error, you can do something about it. With the active approach, it's very difficult to do something about it. In a passive system you have access to the beams, access to the polarization, access to the frequency, etc.

What are the solutions to the Rayleigh backscattering problem?

In a fiber interferometer you use a light source with a lousy coherence, which means a broadband source. A superluminescent diode has been used by many people to get rid of the effects of Rayleigh backscattering.

Will the broadband introduce other problems?

No. In fact, it also gets rid of the nonlinear index effects. It's very good; its kills two birds with one stone. In the fiber resonator case you can't do that because you have to use a narrowband source. You can get rid of backscattering by phase modulating one of the beams before it enters the resonator. Another way is to use different modulation schemes for each propagation direction. This way we keep them apart. We've been working on this for quite awhile, to try to come up with a modulation scheme that won't introduce intensity modulation as well as frequency modulation. I think the effects of backscattering can be reduced quite a lot.

Do some of the other technologies which are being developed these days have a possible role in your projects? For instance, the integrated optical techniques, where you're dealing with waveguides.

Yes. For example, it would be nice to have very-low-loss waveguides. In fact, drawing from the preform into the final fiber smoothes the edges, giving low-scattering loss. But in a waveguide it's difficult to get low-scattering loss.

Every little defect gets propagated.

That's right—edges, homogeneity of the material. It's hard.

Even if you're coupled to a hybrid system?

That's not so bad because in the hybrid scheme you don't need superlow-loss waveguides. You can tolerate some losses. When you make the resonator, that's when you need low loss.

But you could have your laser or your source, your Bragg cell, polarizer, and whatever else you're using on a chip.

The trouble there, again, is what sort of loss you experience when you go from waveguide to fiber. But again there's a lot of work going on in various places on that problem. It has to be solved for optical communication purposes. That's what is going to happen. You will have a substrate with a laser and couplers and modulators, which would then couple into the fibers. So that problem will be solved for communication purposes.

Then you carry it right over to gyroscope applications.

Right. It's funny. In the development of polarization-preserving fiber it was the other way around. It was the fiber gyroscope that pushed the development of polarization-preserving fiber, not communication. Communication is considering polarization-preserving fibers for fifth generation communication—fiber optic communication. For heterodyne communication. For gyroscopes we need that kind of fiber right now—for the first generation gyroscopes, not fourth or fifth generation.

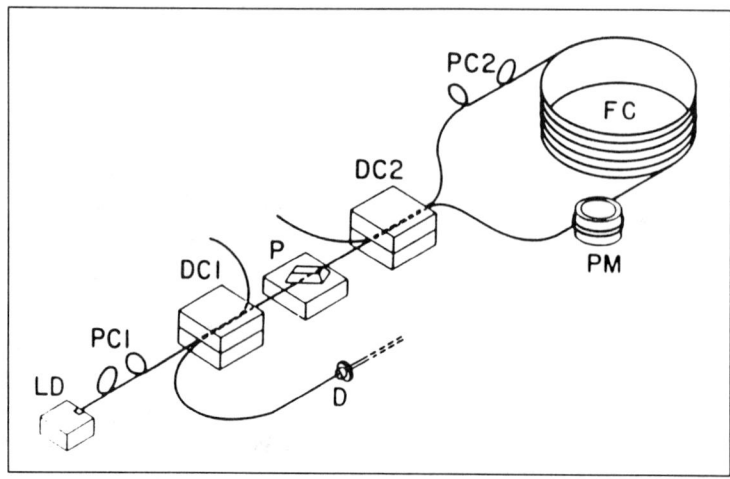

A schematic diagram of an all-fiber-optic gyroscope. LD: laser diode, PC: polarization controller, DC: directional coupler, P: polarizer; PM: phase modulator, FC: fiber coil; D: detector.

So there's a synergism at work in this.

Right, in two different fields. In fact, I think, a lot of communication people right now are interested in the gyroscope.

You mean for the technology it offers for the gyroscope itself?

Because it's a good application for some of their technology, especially in the area of waveguides and waveguide components, e.g., Bragg cells using surface acoustic waves. I think it's a very exciting field. And let's not forget, there are all kinds of other sensors and other uses of fiber optics being uncovered every day. Another interesting area is in semiconductor lasers pertaining to wavelength stabilization. I bring that up because scale factors of all these sensors depend on the wavelength. If the wavelength varies, you're in trouble. For narrowband lasers it doesn't seem to be that difficult to stabilize the wavelength, but it would be nice to make the stabilization scheme compact and cheap. The stabilization of the broadband superradiant diodes, I think, would be more of a challenge.

Are people working on that?

Yes. A lot of work has been done on the stabilization of narrowband lasers, but I'm now starting a project to work on stabilization of broadband lasers. The question is, what do you stabilize to? It's such a huge bandwidth.

Why do you consider it so important? What is the technical problem?

It's important because the fiber gyroscope's scale factor depends on the wavelength.

What do you mean by scale factor?

When you have a rotation, it is measured in terms of a phase shift, which depends on the wavelength of light. If the wavelength varies, you think you are rotating. So instead of landing in L.A., you land in Hawaii because of a wavelength change. The other interesting thing about semiconductor lasers is their linewidth. There's a lot of work going on now in many places on the theory of the laser linewidth. Gas lasers are relatively simple, although their linewidth is so narrow that nobody has really studied it very carefully. But the semiconductor laser's linewidth can be very broad due to all sorts of reasons and will be fun to study.

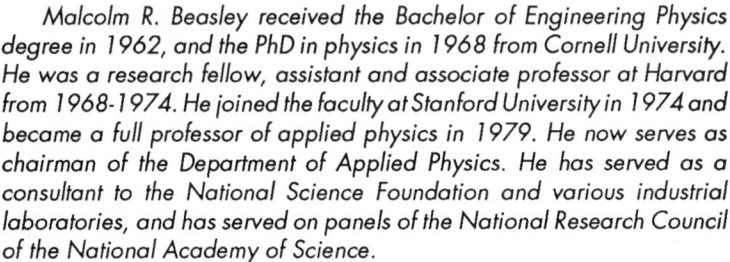

Malcolm Beasley

Progress in Superconductivity

Malcolm R. Beasley received the Bachelor of Engineering Physics degree in 1962, and the PhD in physics in 1968 from Cornell University. He was a research fellow, assistant and associate professor at Harvard from 1968-1974. He joined the faculty at Stanford University in 1974 and became a full professor of applied physics in 1979. He now serves as chairman of the Department of Applied Physics. He has served as a consultant to the National Science Foundation and various industrial laboratories, and has served on panels of the National Research Council of the National Academy of Science.

Prof. Beasley's general research interests are in low-temperature condensed matter physics. Much of his work involves pure and applied research in superconductivity.

Let's start from the beginning. What is a superconductor?

A superconductor is exactly what it would seem to be from its name. It's a material which shows no resistance to electrical current flow. It's therefore a super conductor. To put it into practical terms, if you put a current through a superconductor and measure the voltage across it, you don't see any voltage; it's equal to zero. Correspondingly, if you were to make a loop of material and induce a current in that loop in a closed circuit, and those currents are persistent, it would last virtually forever. So it really is zero resistance. That is not all a superconductor is or can do, but that is the most basic sort of description of a superconductor. If you look into it in more detail, there are other features that a superconductor must have. For example, it must show the Meissner effect, or complete dimagnetism. This means that it will spontaneously expel magnetic fields to make the magnetic induction zero on the inside. If you take a superconductor and put it in a magnetic field above the transition temperature, the magnetic field lines penetrate. If you cool it down, those magnetic field lines come out leading to zero magnetic induction.

Zero magnetic field on the inside of the superconductor?

Right, at least at low fields. High fields it's different, but at least at low fields it will do that, and that is a necessary property that a superconductor must have. It's called the Meissner effect or perfect dimagnetism.

I've looked at some of your notes from the class you gave. In terms

---

Reprinted from *OE Reports*, June 1988. Interview by Frederick Su *of solid state theory and conduction bands, what is the main difference between a superconductor and a regular semiconductor?*

First of all, a superconductor starts out as a metal rather than a semiconductor. Instead of the electrons scattering near the Fermi energy to make normal resistance, in a superconductor the electrons condense into a new state where all the electrons are paired up into so-called Cooper pairs—you have a condensation of the Cooper pairs into the superconducting state, more or less. In that case there's no resistance. Along with that, there's a binding energy between the pairs, and so therefore to create single electrons, i.e., normal-like electrons, there's a finite energy required to break a pair apart. That leads to an energy gap in the excitation spectrum. Thus you have an energy gap in some sense like you do in a semiconductor, but it has a very different physical interpretation. It's not a band gap, it's a gap in the excitation spectrum of the superconductor, with very different physical origins.

Another similarity is that in both cases you can view the conduction as arising from the parallel flow of two kinds of carriers. In a semiconductor the carriers are electrons and holes; and they obey very similar equations of motion. Their properties are very similar; it's just that there are two kinds of them. That distinction is very important, say in a p-n junction, but not so important in, say, the conductivity through a bulk material. In a superconductor, on the other hand, there are also two kinds of carriers: the normal electrons and the superconducting electrons, or Cooper pairs. But in the superconducting case, the superconducting electrons have totally different behaviors than the normal ones. You have very different circuit properties, rf properties, as a consequence.

Examples of such behavior?

Examples would be that the surface resistance of a superconductor goes exponentially to zero at absolute zero temperatures, whereas for a semiconductor you have finite surface resistances. Basically it's the superconducting electrons with their lossless current flow that make the rf properties of superconductors so unusual. It would make transmission lines lossless and dispersionless. You can't do that with semiconductors because the carriers always behave like normal electrons; they have resistance, and therefore do not have such nice properties. So conceptually you can think of it as two fluids or two types of carriers. But the superconducting pairs behave so differently. That's where all the unique properties of a superconductor come from. If you think of

The surface resistance of a superconductor goes exponentially to zero at absolute zero temperatures, whereas for a semiconductor you have finite surface resistances. Basically it's the superconducting electrons with their lossless current flow that make the rf properties of superconductors so unusual.

a simple equivalent circuit for a superconductor that has the essential physical nature of it built in, then think of a resistor and an inductor in parallel, where the inductor is really a perfect conductor, that is, has no resistance. In that case, inductance arises from kinetic effects, simply from the kinetic energy of the super-conducting electrons. You can understand an awful lot of the properties of superconductors from such a simple picture.

Can you give me a really brief history of what led to high-temperature superconductivity?

Historically there's been a long tradition of seeking new materials in research in superconductivity, and there are a number of people who play that game. Many people have tried. The principals in the recent, new high-T_c superconductors were first Bednorz and Mueller, and then Paul Chu. Bednorz and Mueller really drew our attention to a new class of materials, the layered perovskites. Then Chu found a very fortunate chemical modification that pushed the transition temperature up to 90 K. It's up to 125 Kelvin, and it will go higher, I think, even within this basic class of materials. So that's the history—the long, frustrating search finally paid off, and somebody found not just a better superconductor within known classes, but really a new class of materials on which the transition temperature has really shot up. It's as much of a revolution scientifically as it is technologically. Scientifically, these materials are raising fundamental questions which permeate broadly

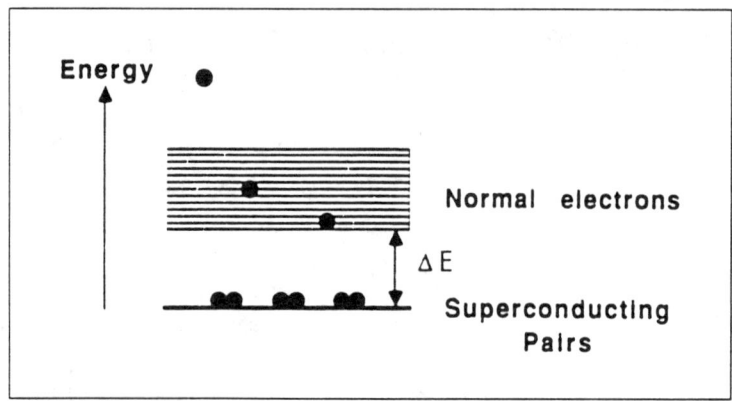

In a conventional superconductor the electrons form superconducting pairs with a binding energy $2\Delta E$. Since these pairs act like bosons, they can all exist in a single quantum state. Excitations from this state act like normal electrons. Hence there is an energy gap in the energy levels of a superconductor. The density of normal electron states has a gap with a peak in the density of states just above the gap.

throughout condensed matter physics. Technologically, the potential of these materials, if useful electronic or electrical applications can be found, is equally profound. So it really is a major breakthrough.

When you're talking about technologically, you can talk about electric transmission lines and power loss through Joule heating. Superconductivity gets rid of that kind of thing, right?

No, it does not eliminate all losses in power lines; the field levels are too high and hysteretic losses arise. But it does reduce them, yes—and perhaps even to a useful level according to recent estimates made at EPRI by Mario Rabinowitz.

What about when you're talking about electronics?

Let's consider first where we were before the new materials. There are, after all, a lot of highly successful superconducting electronic devices. SQUID (Superconducting Quantum Interference Device) detectors and instruments, and high-speed Josephson logic all perform very well. They've not had a major impact, I think, for two reasons. One is the inconvenience of lower-temperature operations, cryogenics and that sort of thing. The other is that to make them into larger systems, to incorporate them in semiconductor systems, was not possible. You had to have the entire system superconducting, and that meant that you had to make every component superconducting. That's hard to do, and expensive to develop, and so on. It has never been done for a large system.

The new materials change both of those factors rather dramatically. It is now possible, if we can get good electronic behavior from these new materials, to do the old things that superconductivity can already do, but do them at higher temperatures. That will broaden the range of applications where people will be willing to consider them, because the cryogenics will be so much simpler. More profound, perhaps, in the long run, but certainly more difficult to assess, is that now the temperature range over which semiconductor devices and superconducting devices work overlap. And it is not just semiconductor devices. It's whole systems. You can imagine hybrid systems, hybrid devices; you can imagine hybrids at the physics level where you get more intimate in the connection. Nobody has really thought any of this through. It seems to me that all this profoundly changes the technological equation.

Under what conditions is optimal use for high-T_c superconducting devices defined? Semiconductor devices?

That's tough to say. It's going to take a while to sort out. The one place where the sorting out is currently going on now, I would say, is in the area of interconnects—taking these new superconductors and making wires or transmission lines out of them and connecting up semiconducting devices. Setting aside the materials problems and such matters, that's a question one can pose now. And it is being addressed on the assumption that the new superconductors will provide transmission of comparable performance to those that we have with conventional superconductors. From what I hear, it appears clear at this point that any major impact would require a different computer architecture. Just replacing present metal interconnects with zero resistance superconductors probably doesn't buy you much. After all, these are already well optimized systems. But if you were to take advantage of the fact that you had transmission lines that could go long distances with no dispersion and no loss, you might organize your system very differently and get major gains and performance from the system. Nobody has proved that yet, but I think that's the kind of thing people are looking at very seriously.

That's a question I've always had. Conventionally, what is the power loss and the dispersion for electric transmission lines now?

Well, for signal transmission lines, over distances you would go in a computer—say, ten centimeters or so—the loss is, as I understand it, all in the dielectric. You have virtually lossless, dispersionless transmission lines. Picosecond pulses—i.e., few picoseconds to 60 ps pulses—don't decay and they don't broaden. This has been demonstrated with conventional superconductors.

What about the electric power transmission lines over distance?

In signal transmission lines—striplines, microstrips—basically the superconductor always stays in the Meissner state. That means the fields do not penetrate the superconductor, and you have these very nice propagation characteristics. In power transmission lines there are magnetic fields on a superconductor of a magnitude such that the field penetrates, that is to say the Meissner state breaks down. Under these conditions the superconductor is in the so-called mixed, or vortex state. When this happens you don't have the ideal properties of the Meissner state. You have hysteretic losses, generically not unlike those associated with the B-H curve of a ferromagnet.

You have hysteretic losses, and those losses are not small under all instances. So you have to optimize the material for power line applications so as to minimize hysteretic losses. Optimization for these two classes of applications is very different. In one case, you're trying to optimize the Meissner behavior; and in another case you're trying to minimize hysteretic losses.

Getting back to your domain, you said there's a crossover between the high-temperature superconductors and semiconductor technology. . .

An overlap of the temperature range.

And you were talking about interconnects too. What about using the high-T_c superconducting materials for detectors in optoelectronics and optics?

Well, that's an interesting possibility. I think we don't really know. There's been a lot of interest in such devices. In conventional superconductors we know that you can make very-high-frequency mixers, I mean up into the millimeter and submillimeter wave band. There's every reason to think that we'll get up into the terahertz range; that's not quite infrared but it's getting close. There is, of course, an energy gap in superconductors, and one can imagine devices based on direct absorption. These break pairs and create nonequilibrium excitations in which you make normal-like electrons. Then, in principle, we could make a detector if one could detect the presence of those normal electrons somehow. I don't think anyone has done this yet, at least not in an effective way. But you could imagine it.

I think much of the interest in infrared detectors is in the micron range. There was work in Japan claiming that granular lead-bismuth-barium oxide superconductors had good infrared detection properties. Now these new superconductors are not unlike that original material. That was a perovskite too, by the way. It's an interesting question whether that old perovskite superconductor was really a precursor of the new ones or not. We don't really know. But in any event, NTT in Japan made films of lead-bismuth-barium oxide and showed that it had infrared detecting capability. Now the numbers they quote are outstanding, but there are some critics. We'll have to wait and see. Certainly people are now working hard to clarify the situation, and to ascertain whether it's just an effect, or something usable for technological purposes. Now I think there's also a possible role for superconductors in focal plane arrays. Again, I'm not an expert, but as I understand it, if you have focal plane arrays of infrared detectors of any sort, you have a data processing problem handling all the data. From what I hear, there is some interest in using the very-high-speed computational capability of superconductors to handle the data flow and to do signal processing right down there at the array. After all, you're already at low temperature. So there's interest in taking superconducting digital signal processing circuits just to handle the data that would

I think it's accepted that you can make digital signal processors with superconducting electronics. It's a question of where you have the need to develop it.

Malcolm Beasley   **113**

be generated by these arrays.

Would superconducting electronics have faster switching times too?

Oh yes. The whole IBM program demonstrated rather impressive computational capability on the part of superconducting logic. They were very fast, and it all worked. The difficulties lay more in

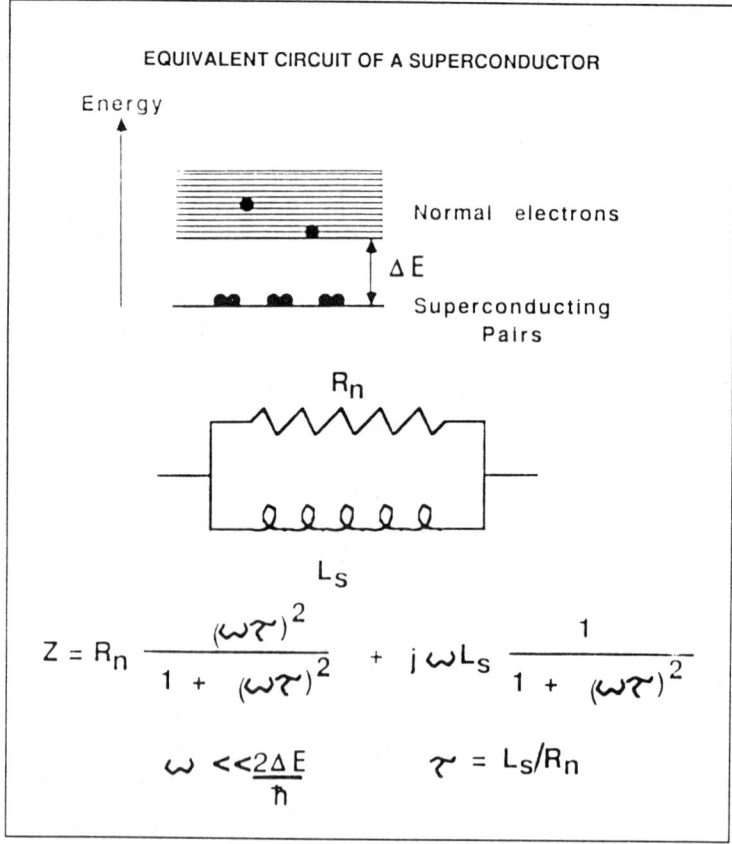

EQUIVALENT CIRCUIT OF A SUPERCONDUCTOR

There are two kinds of carriers in a superconductor, the superconducting pairs and the normal electrons. The pairs have no resistance and act electrically like a perfect inductor. The inductance arises due to the kinetic energy of the pairs, and is the so-called kinetic inductance. The normal electrons exhibit resistance just as in normal metals. The electrical equivalent circuit of a superconductor hence is a perfect inductor in parallel with a resistor. The frequency responses follow from this. The most important result is that the resistance is zero only at dc and rises as the frequency squared for finite frequency. In practice, these losses are very small, zero at zero temperature where all the normal electrons are frozen out. It is the purely inductive response of the superconductor that leads to its outstanding performance in signal transmission lines.

memory, and in the economics for their particular interest as I understand it. I think it's accepted that you can make digital signal processors with superconducting electronics. It's a question of where you have the need to develop it.

Previously everything was liquid helium temperatures for supercon-ductivity, and that was 4 degrees Kelvin. Where are we now and what of the future?

125 degrees Kelvin for sure, and I think it will go up more. Many people believe they know what to do to make it go up more. I am inclined to agree. I'll even make a prediction—one I believe to be quite responsible. I think it is extremely likely that we'll see the transition temperature go up to something like 140, certainly about 125, which already exists. This means that operating at liquid nitrogen temperatures will surely be possible. Liquid nitrogen is 77 degrees Kelvin, and thus you're at less than half T_c. Therefore, you're quite comfortable. It was not so comfortable to imagine 77 Kelvin operation for a 90 Kelvin superconductor. You would be awfully close to the edge in that case. The interesting question now is whether they will provide the devices and device functions and all the other things you need—there will be formidable materials problems—and on and on. But to the basic question of whether the T_c is high enough, the answer is yes.

Can you give me specific examples of some of those thallium compounds?

Well, there's the thallium-barium-copper oxide family of com-pounds (chemical systems), which has transition temperatures as high as 125 Kelvin. That's relatively new, but it's been confirmed; it does exist.

What's the best guess as to the properties of the new materials that lend themselves to superconductivity at high temperatures?

Who knows, who knows? That's the big question from the physics point of view. I think the most solid conclusion that you can draw is that it isn't like in the old superconductors. In the old superconductors where we know the answer, and in most in-stances we do, the superconductivity arose from an attractive interaction between the electrons that arises from what's called the electron-phonon interaction. In physical terms, as an electron moves through the crystal, the crystal ions move in response to that electron. It leaves a positive attractive potential locally. If you like, you've got a medium that's polarized by the electrons, and it's that residual polarization left in the lattice that attracts the second electron. So it's indirect, via the lattice. Imagine a rubber sheet, and some particle comes along and makes a dip in the rubber sheet, and then goes away. Then another particle comes along, sees that dip, and is attracted to it. Now, in effect, that makes an attraction between those two electrons. That's kind of a simple-minded view of electron-phonon interaction. Now in these new superconductors, nobody knows whether there is a new kind of phenomenon here. That is to say, this is a wholly new way of making superconductors, or whether it's like the old ones, but we need a different method to attract the electrons. Some people think it's due to magnetism and magnetic interactions between the electrons. Some people think it's so-called exitonic, where you have very strong electronic polarizations, that is, polarizations of the electrons, not the ions in the lattice, to make the attractive potential. There are some people who have a lot more exotic interpretations. It's really just not known. I think everybody agrees that there's a new secret of nature that will be revealed here, that the old understanding will not be sufficient. Most people believe that a real new picture will come forth when we finally understand it.

I have a feeling that whoever understands it first will get the Nobel Prize.

It depends on what the answer is. I wouldn't rule it out, but I don't want to say it's a sure thing; I think it depends on how profound the answer is, and how wide its impact in our understanding of physics. The BCS theory, which was awarded the Nobel prize, not only explains superconductivity, but it had a conceptual or theoretical content that impacted on physics theory very broadly. All kinds of physics were better understood when we understood the theory of superconductivity. One hopes that will happen again, but you don't know. I certainly hope so.

K. E. Creer

Laser Detection

Scotland Yard's High Tech Crime Fighting Techniques

K. E. Creer is head of the Serious Crimes Unit at Scotland Yard's Forensic Science Laboratory. He joined New Scotland Yard as a crime scenes photographer in 1961 and spent seven years covering many aspects of police photography. In 1968 he moved to the Forensic Science Laboratory to set up a photographic section; previously the photography had been done in a limited way by the scientists themselves. During the next 12 years the section built up a worldwide reputation in forensic photography. The acquisition of a 2-watt argon-ion laser in 1980 led to a heavy involvement in the development of latent fingerprints.

In 1983, Creer was awarded the Fellowship of the British Institute of Professional Photography for his thesis on forensic photography with lasers. In 1984 he received the Applied Photographer of the Year award for his work with lasers. In 1986 he was awarded the Fellowship of the Royal Photographic Society for research into the detection of trace contact evidence by laser excited fluorescence. He has published 22 papers on many aspects of forensic photography, and has lectured at many conferences in several different countries.

Perhaps you can tell me about your background and how it led you into your line of work.

I'm a photographer by trade, not a scientist. I started 26 years ago at New Scotland Yard as an operational photographer, visiting scenes of crimes. In 1968 I moved to the forensic science lab to set up a scientific photography unit. We'd gotten involved in ultraviolet, fluorescence, and various other types of lighting. I was also very interested in lasers. When the papers started coming out in the mid and late 70s from Roland Menzel—he was using lasers to detect fingerprints—we got very interested in this. In 1980, we purchased a 2-watt argon-ion laser. Very quickly, we got excellent results. I think we had a bit of luck: within two weeks we had a big armed robbery where the security box that was snatched had an alarm triggered and sprayed orange dye around. The robbers dumped the box in the road, shot the lock off, took the money, and ran. The box

Reprinted from *OE Reports*, January 1988. Interview by Roy Potter.

came in to me; we couldn't see any fingerprints, but the orange dye fluoresced strongly. We turned up the criminals' marks in the fingerprint collection of Scotland Yard, and they were arrested. So this got us off to a good start and everyone started to say "The laser's got some applications." We spent the next two or three years building up the operational side. Obviously the argon-ion laser isn't portable, so items had to be brought back. We found it was lacking in power, so in '83, we bought a 12-watt argon-ion laser from Spectra-Physics with a UV option, as well as a dye laser. We've got rhodamine 6G in the dye laser, giving us 585 to about 620 nm. I've found that by working in a single spectral line, you've got more of a chance of avoiding background fluorescence. The big problem in our work is that we don't know what substances we're dealing with. Criminals don't wash their hands before they commit crimes, so they leave finger marks in all sorts of contaminants, as well as in the sweat. Because we don't know where these contaminants absorb, we don't know where they emit. Ideally, we'd like an infinitely tunable laser, and so would everybody else! The nearest we could get was using the eight single lines of the argon-ion laser in the blue and the green, the dye laser, and the UV option, so that we could get long-wave UV. If the case is important enough, we'll examine every wavelength if necessary, hoping for the best. It's a bit hit-and-miss, but we've had some excellent results. You have to look very, very closely. I may spend an hour examining one super-market plastic carrier bag, for instance. So it's a long, time-consuming business.

If you're looking at a plastic bag, what are you looking for?

Fingerprints. We find plastic bags are quite a good surface to get fluorescent fingerprints. They're used in a lot of crime; criminals carry guns and stolen property in them, so they turn up frequently.

And if you carry them around, they're rather innocuous.

That's right.

When you lift prints, would it be possible to pick up prints with techniques other than fluorescence?

Yes, indeed. We've got a number of techniques, and the laser is just one. We find the laser is useful, but we're also interested in making prints fluoresce by using chemicals that either tag onto the prints or react with them. The standard one, which has been used widely in the States—we use it also—is to fume the exhibit with Superglue fumes. You put it in a chamber, the Superglue forms a polymer reacting with the fingerprint moisture, we think. They're quite difficult to see because the prints are either transparent or white. We then stain with rhodamine 6G in methanol. The rhodamine gets trapped in the polymer, the rest is then washed off, and you're left with fluorescing prints. We get some quite good results with that. There's a whole range of other techniques. What's happened is we've gone a bit outside my discipline, which is photography and light, so we have two chemists in the unit, along with two fingerprint experts. The unit in 1984 became known as the Serious Crimes Unit, dedicated to applying all available techniques to get evidence. It's staffed now by scientific photographers, chemists, and fingerprint experts—only a small group.

The professional members of your laboratory are also trained to appear as expert witnesses.

Yes, we spend quite a lot of time in the witness box. Within the Serious Crimes Unit, we are finding we are being challenged on fluorescent fingerprints, because the defense can't see them. Of course we use the laser as a first line of examination. In a murder case, we will examine the murder weapon, and other important exhibits, and subject them to a range of treatments—we have 22 treatments we can use—and in some cases we can sequence four or five treatments.

What are some of them?

I think paper would be a good example. If we had a kidnapping, we would bring in the ransom note and examine it with the argon laser using the blue and green lines, the dye laser, and ultraviolet radiation. We'd also use the dye laser and follow everything with a range of chemical treatments.

Are you looking for a reflection from it?

No, fluorescence. We run the laser beam down a 5-mm liquid light guide that diverges the beam; obviously you can't put a laser beam straight onto paper. We've got a series of safety goggles with sharp cutoff filters ranging from 530 to 630 nanometers, so we can change goggles. What we're looking to find is a wavelength where the fingerprints fluoresce and the background doesn't, or where we can subdue background fluorescence. We supplement these goggles with a number of interference filters; they're 7-nm half bandwidths.

To separate the fluorescence?

Yes, if you've got background and mark fluorescence, you can find that window which transmits the fingerprint and cuts out the background. Paper does give us a problem, especially using blue and ultraviolet radiation; the whiteners in the paper fluoresce. We carry out a thorough laser examination, photograph anything we

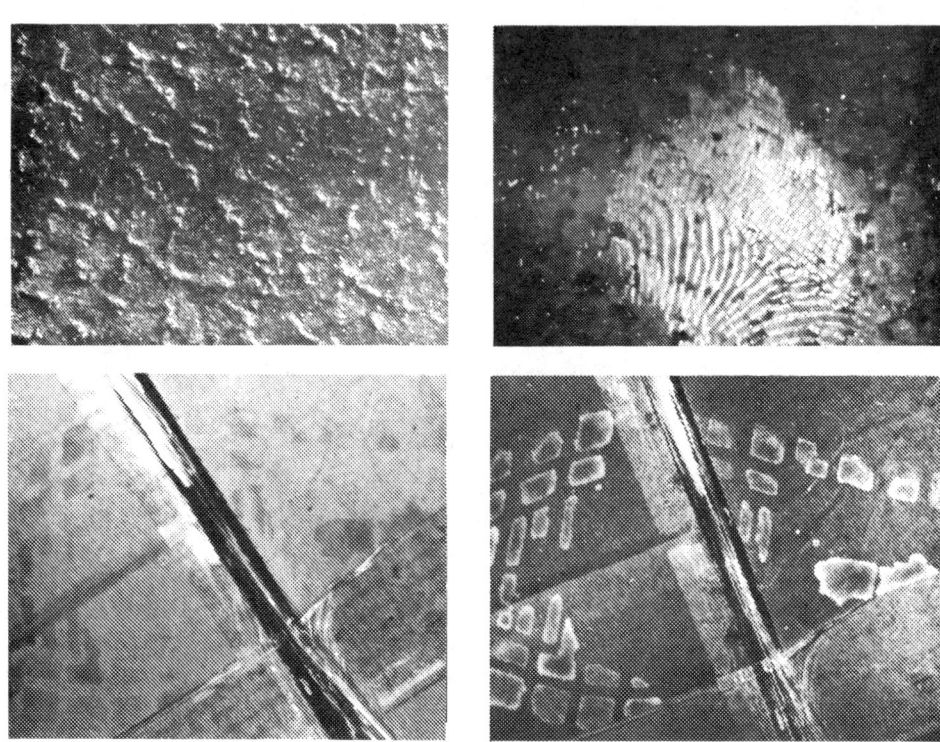

Photographs from K. E. Creer's paper in SPIE Volume 743, Fluorescence Detection. Top left to right: fluorescent fingermark on stippled surface. Next: fluorescent shoeprint on cardboard. At right: Fingermark (invisible) under normal light, and (below) through 550 nm, 720 nm, and 850 nm filters.

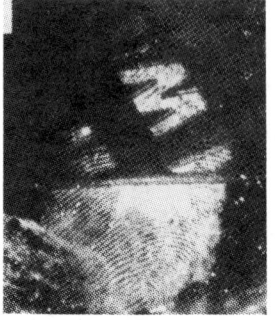

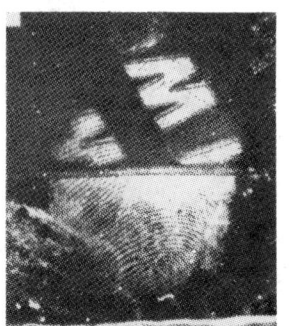

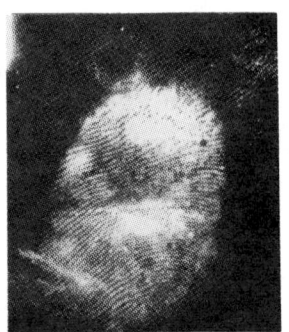

get. To photograph marks, we put a similar filter over the camera lens, and paint with light in total darkness with the camera shutter open. Exposures can range from a few seconds to a few hours in some cases, depending on the amount of fluorescence. With paper, we would follow this with conventional ninhydrin treatment. It's a chemical which reacts with the amino acids and produces a stain, rhumans purple. It's a very effective method that's been used for 20 or 30 years now, and it produces a lot of finger marks, which can then be photographed. If you produce very faint ninhydrin marks with insufficient detail for the fingerprint expert to identify, we spray with zinc chloride. This reacts with the rhumans purple and forms a zinc ninhydrin complex, which fluoresces under the laser. We've found that if we freeze it in liquid nitrogen, we can increase the fluorescence tremendously. We use a polystyrene tray and immerse the paper in the liquid nitrogen. This produces fluorescing marks that you won't see at room temperature. We then photograph them while under liquid nitrogen. This causes some problems with condensation on the camera lens, but we do get some excellent marks that way.

Do the vapors from liquid nitrogen get in the way?

No, I've found the vapor from liquid nitrogen, being heavier than air, drops out of the sight of your box. The fluorescence is so bright that you need a fairly short photographic exposure, and we have three or four minutes before everything mists up. We follow this with a treatment known as physical developer; this is a silver nitrate-based treatment where the silver deposits out of the lipids, the greasy parts of the fingerprints. This reacts with a totally different chemical in the fingerprints.

Let's say you find some marks early in these processes. Would you still continue because you might find other marks as well?

Oh, yes. Each stage can give you quite different marks. The laser produces marks that don't react with the chemicals, the ninhydrin can develop the amino acids which may be there. Different people produce different amounts of amino acids. The physical developer will react with the lipids. The good thing about that is that when the paper has become wet, which sometimes happens, you lose the water-soluble amino acids, while the lipids stay. We've had paper submerged in water for several days, and still developed fingerprints on them with physical developer. We then meet another problem: if the fingerprint is over newspaper, the physical developer develops a very dark silver image, which is impossible to separate photographically from the printing. Then we use the scanning electron microscope, equipped with an extra-long chamber, since the normal magnification of the microscope is too high for a fingerprint; this enables us to record the silver and lose the printing. In a very serious case, we'd carry out all those treatments just on one sheet of paper.

And so that one piece of paper could have been subjected to a half-dozen or so discrete tests.

That's quite possible. I had one case where we produced a laser fluorescent print. At the physical developer stage we got another print that was in the identical position; they were on top of each other, but because the fluorescent print didn't develop with physical developer, and vice versa, we got identifications for two different people. They were both involved in the crime.

In the mystery movies they always have the criminals wiping off their fingerprints. Have you overcome that problem?

No, if it's wiped thoroughly, you'd probably lose them. We've progressed from bringing items back to the laboratory. There were increasing demands for us to attend the scene, because a lot of items, like walls, can't be brought back. So we set up an experiment in '85. The British firm, Spectron Lasers, built us a Nd:YAG laser, frequency doubled, to give us 532 nm. We also had it frequency tripled and quadrupled to give us long wave and short wave UV. Unfortunately it was about the size of a desk! During an experimental period I went with a research scientist from our Home Office who specialized in fingerprints to a number of major crimes in London. I took the YAG along with infrared image converters and a UV light. I illuminated the scene while he was looking at developing evidence with chemicals. We were reasonably successful, but one thing we did learn is that it's very time-consuming. At the end of this experiment we decided to continue it. We had enormous problems carrying this laser to a scene; it took four of us to get it in and out of a van. If a murder happened two or three floors up in a tenement block with no lift, we couldn't get it into the scene. Quite apart from fingerprints at the scene, we were finding fibers on bodies that sometimes could be useful. A lot of fibers that are very difficult to see in normal light will fluoresce. Occasionally we found fluorescent shoe prints. I found that bruises and similar marks on bodies could be enhanced; so in a strangulation case, you could actually bring out the texture of the rope or whatever was used to strangle the person. The skin fluoresces and the area of damage absorbs, so you can get a big contrast enhancement.

Does that change with time?

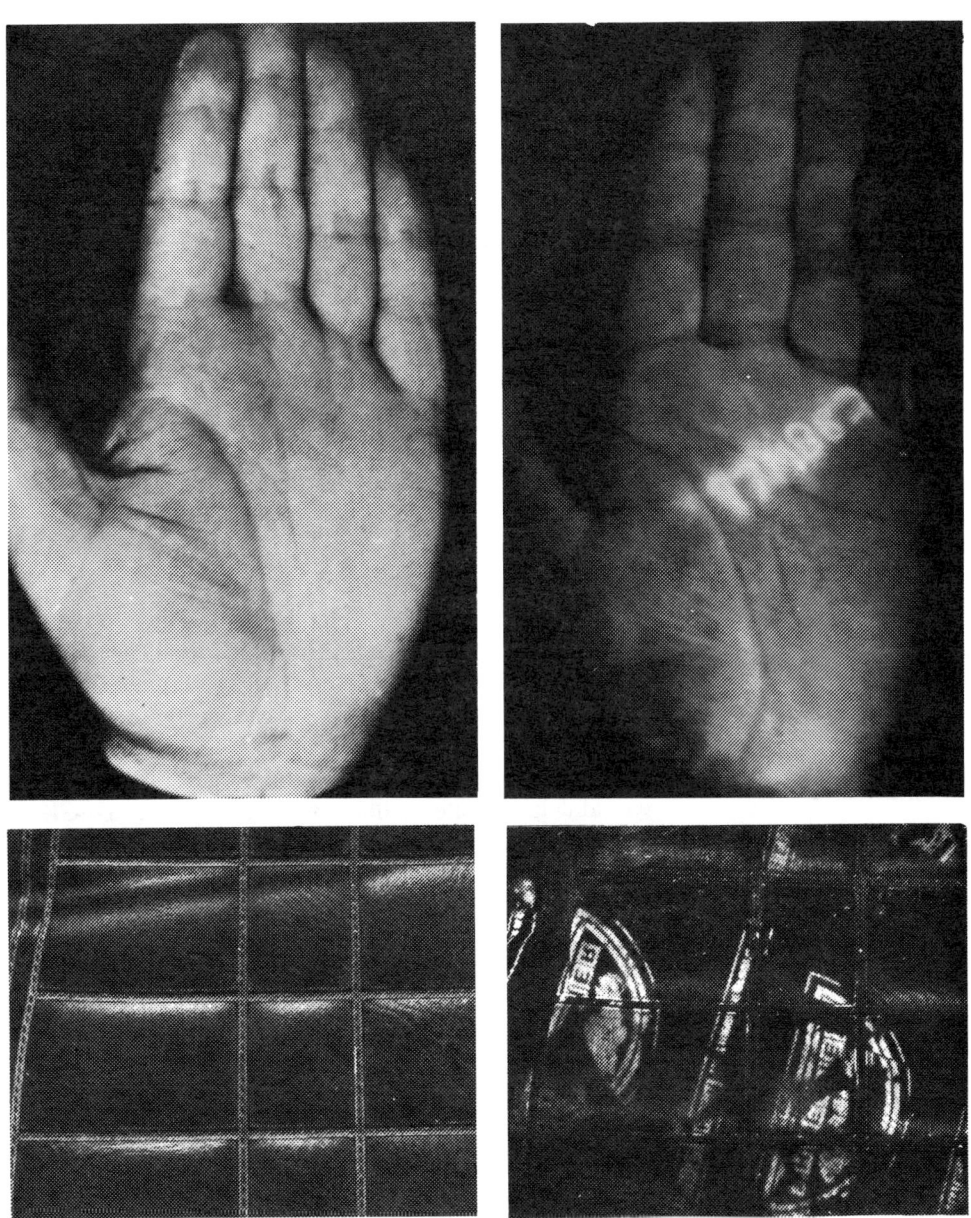

Top: Recovery of writing on hand (normal light/laser). Bottom: Transfer of printing from carrier bag to motorcycle saddle.

It improves over two or three days, usually, as the bruising starts to come out of the dead body. So we will look immediately, and quite often we will go back two or three days later and look again. We've had some excellent results. So we've now formed a fully operational unit—I'm on call 24 hours a day with the police. We only go to the big murders.

What's a big murder?

What we call the whodunits. In a husband-wife argument, most of the evidence isn't much use because they both live there anyway. But in cases where a criminal has broken in and killed someone, we try to get the police to preserve the scene. We get there as soon as possible and search the entire scene with light; it can take a long time. I spent 23 hours searching a house last year with a small laser beam, throwing a beam about three or four centimeters in diameter. We do that first, then the teams move in with the conventional methods.

Are these green lines you're using?

Well, when we were using the YAG laser, we were using the 532 line. Last summer a firm from California called Omnichrome came over to London to demonstrate a 200-mW argon-ion laser, very nicely packaged—it's about the size of a sewing machine, under 50 pounds. They loaned it to me for operational trials, and the first two murders we went to, we got a lot of marks that we wouldn't have gotten otherwise. It's a lot easier to use than the big YAG laser, and of course being CW, it's a lot easier on the eyes. We operated the YAG laser at a repetition rate of about 24 Hertz, which is very uncomfortable. We've now ordered the argon-ion laser. You've got to look carefully, and you've got to blacken the room, but 200 mW will find you plenty of marks. Their delivery system is by a single fiber, so it's pretty efficient. It has a small lens on the end that diverges the beam.

How does the defense handle a challenge to your evidence?

Well, the defense has their own experts, and they are quite entitled to come up to our laboratory, and we offer them the use of the facilities. They can check what we've done. We've got nothing to hide, so we work quite well with the defense. The normal attack in court is not so much on the facts of the case; it's more on what we call the continuity. If the exhibit is brought back to us, they want to know what time it arrived, who handed it to you, did you hand it back; and if they can get someone to say something different, they can cast doubt over what happened to that exhibit. If it's declared inadmissible, we lose all the evidence. So the

> **Some bank robbers dropped a sports bag in the bank. While I was searching it, up came a very clear name on the bag. Obviously the criminal had had his name on the bag and had used a solvent to take it off. He was very unlucky, because we weren't even looking for his name, we were looking for fingerprints.**

challenge is on the periphery, not on the facts. We have to keep very, very accurate records of what happens to the exhibits, because that seems to be a standard challenge.

Do you think your techniques have improved the ability of the police to solve a number of crimes?

Oh, yes. The big problem is that they're so time-consuming. We're only able to apply them to a tiny fraction of the most serious crimes, and they require a fair amount of enthusiasm and skills to make them work. We've been looking at ways that we could make laser detection of evidence available to our crime officers at the scene, who go out to all the crime scenes. It's very expensive to equip them with lasers to go to only routine crimes—burglaries, things like this—and find evidence. We obviously could not get involved in this, it's so vast. We only deal with the top level.

I would expect you have a lot of interaction with other national police departments.

Yes. We had visits from the German police, the Italians, the Dutch. There's been a lot of interest in it, but nobody else in Europe has yet set up a unit using lasers. We've also got very close contacts with the FBI and the Canadians. Lasers have been used much more widely for forensic purposes in the United States than anywhere else in the world. The FBI has an excellent unit, a lot of your police departments have units. We meet a lot of Americans who come over to see what we're doing. I've published quite a number of papers over the years, as has Roland Menzel, who started all this off. He knows far more than I ever will about how fluorescence works, about how lasers work, but he hasn't got the interest I have in actually detecting evidence. I take his ideas, if you like, and make them work. I've done two or three programs with Roland, and we complement each other nicely; he talks about the theory, and I talk about how it works. That's what we're doing this afternoon, really.

What do you see as some problems that you'd like to attack in the near future?

One thing we're starting to do at major crime scenes is to examine the scene with other light sources. I've recently purchased a 200-watt high-pressure mercury vapor lamp. It's an inspection lamp for metal defects. We're having filters put onto the front of the liquid light guide to isolate each of the mercury lines, so we can get the 365, the 435, the 546, and the 578 nm lines; by using each of these lines, I'm hopeful that we'll find more evidence. I'd like a bigger mercury vapor lamp, but you soon start moving into problems with cooling and bulk. At this exhibition I've been talking to a firm about a 1000-watt mercury vapor lamp, but I'm not sure the filters would then stand up to the heat. What we're doing is putting interference filters over the end of the liquid light guide to isolate each line, and we've got to isolate it completely, because the fluorescence may fall where one of the other lines is. If you don't cut that out, you won't see fluorescence.

You indicated you're using infrared imaging too. How does that fit in?

We've got an infrared video camera with a silicon vidicon tube. I think its peak is about 820, so it's just out of the visible. We occasionally find fluorescent fingerprints, particularly if you excite with a dye laser at 585, the Stokes shift will put the fluorescence out into the far red and near infrared. It's very difficult to search by a video system because I find, to resolve a fingerprint, you've got to put it on the screen. If you move your camera quickly, you get retention problems; faint marks are very difficult to resolve.

Can you give some examples of cases where the laser has been particularly useful?

We had an armed robbery where a security guard was shot. A stolen motorcycle was used as the getaway vehicle, and then abandoned. It came into us for the fingerprints. When I put the laser on it, I found a fluorescing pattern on the saddle. We identified it as the motif from a carrier bag from a wine shop. The owner had covered the saddle with the bag when it was wet, to keep his trousers from getting wet. Anyone sitting on that would have a transfer back onto the seat of his trousers, and he wouldn't see a thing. Unfortunately we didn't catch a suspect. We had another one where some bank robbers dropped this Adidas sports bag in the bank. It came in to me, and while I was searching it, up came a very clear name on the bag. Obviously the criminal had had his name on the bag and had used a solvent to take it off. He was very unlucky, because we weren't even looking for his name, we were looking for fingerprints. I phoned the robbery squad, and they knew him. I had another one where we had a hit and run accident involving a Dutch lorry. The number was broadcast to all police cars. Some miles away there was a policeman out of his car talking to someone when he saw a Dutch lorry drive by. He wrote the number down on his hand with a ball point pen and then went to check the number in his car. It wasn't the hit and run lorry, so he forgot about it, washed his hands, and the number was washed off. The next day he heard that there had been a hijacking of a half million pounds' worth of goods in a Dutch lorry. He thought it was the one he'd seen, but there was nothing left on his hand, and he couldn't be sure. We had a very nervous police officer sent up to the lab to put his hand under the laser. The skin fluoresces quite strongly, and with the conventional safety goggles and the laser we couldn't see anything. But when we went down to the 620 nanometer interference filter, I transmitted the writing, held back the fluorescence of the skin, and the number was visible. It was the lorry that had been hijacked. According to the driver, at this time it hadn't been stolen and he hadn't been in that part of Essex where it was seen, so he was obviously lying. It was an insurance fraud. There's a whole range of these sorts of cases where the laser has been of value. From what I can see in the States, the laser is not being used for these purposes—it's mainly being used for fingerprints.

Are you ever called in to identify fake artwork?

Occasionally I do, in a private capacity, as a consultant. If there's no physical crime involved in the case, we're allowed to, in effect, hire the lab from the commissioner of police. We'll pay a fee and come in on a weekend to use the facilities. We obviously only work for reputable companies, and each one has to be approved at a fairly high level. I occasionally do cases for Christie's the Auctioneers, and for Lloyd's, the insurance people. I had a very interesting one for Christie's. An old lady had died in Wales, and they were asked to look at the furniture. They found a desk that they thought might have come from the Palace of Versailles at the time of the French Revolution when the furniture was looted and scattered all over the place. Underneath the desk the wood was just blackened with age, but there was just a trace that there may have been a number there. Apparently in the Louis XVI reign, all items of furniture had lampblack numbers put on them, and an inventory was kept that was still in existence at Versailles. I put the laser on this desk; the wood fluoresced, and the lampblack, or whatever it was in, absorbed. We got back the number; it showed that this desk had been made for the sixth daughter, I think it was, of Louis XVI. It had been valued at about £10,000, and the Palace of Versailles bought it back for a quarter of a million! I just charge a fixed fee; I wish I'd charged a commission. There's quite a potential in the art world for the use of lasers, and as far as I know, nobody's using them. Another possibility that I've never done, but it occurs to me: We know very old marks will fluoresce—that's one of our problems. Now suppose you've got a painting that you're trying to prove is a Gainesborough or whatever. If you could look at a lot of paintings by him, you may actually find his fingerprints on some

Suppose you've got a painting that you're trying to prove is a Gainsborough or whatever. If you could look at a lot of paintings by him, you may actually find his fingerprints on some of them. You could then use those to match up to prints found on anything else. It's quite an exciting thought, and it would be much more positive than the judgment of some art experts.

of them. You could then use those to match up to prints found on anything else. I may be going way out on a limb, but I don't know if anyone's ever tried to apply this. It's quite an exciting thought, and it would be much more positive than the judgement of some art experts.

This has been a fascinating conversation, and I thank you very much.

Jules S. Jaffe

The Role of Optics in Deep-Sea Exploration and the Search for the Titanic

Jules S. Jaffe received a BA in physics from the State University of New York at Buffalo in 1973, an MS in biomedical information science from Georgia Institute of Technology, Atlanta, in 1974, and a PhD in biophysics from the University of California, Berkeley, in 1982.

After spending several years working in industry as an image processing consultant, he joined the Woods Hole Oceanographic Institution as a Pew Memorial Fellow in 1984 at the Assistant Scientist level, and became an Associate Scientist in 1988. There he concentrated on acoustic imaging, underwater light imaging, and satellite imaging of the oceans. This line of research played a dramatic role in the discovery of the Titanic. He is currently an Assistant Research Oceanographer at the Scripps Institution of Oceanography, Marine Physical Laboratory, University of California at San Diego, which he joined in 1988. His research interests are in the areas of image reconstruction and restoration, with a special emphasis on three-dimensional systems. He is currently designing underwater optical and sonar imaging systems for ocean exploration.

This interview was conducted by Roy Potter, SPIE Technical Consultant.

Did you go out on the Titanic excursions?

I was asked to participate in the Titanic voyage, and I elected not to go. The most common question people ask me is "do you regret that?" I suppose I don't, because the North Atlantic is not a very friendly place to be at any time of the year. In fact, after the discovery, while they were still at sea, I was sort of a technical liaison between Woods Hole and the press, and I enjoyed a lot of activity, more than I certainly would have had at sea. However, I did not experience the thrill of discovery. Recently we were on a scientific cruise, and I was at sea for three weeks. We took the Argo system to map the East Pacific Rise, which is an area where the tectonic plates are spreading apart. It's the fastest known spreading center on the surface of the earth. That was quite exciting.

You were looking at the vents?

That's correct. Back in the mid-70s, in the submersible Alvin, Bob Ballard and others participated in a cruise which resulted in a major geological find, probably the most significant biological discovery of the century.

Excerpted from *Optical Engineering Reports*, June 1986.

Alvin is a submersible that carries three people?

Correct. It's a titanium sphere, untethered. You're on your own when you're in Alvin.

How long does it take to get down?

The transit down can take anywhere from two to four hours. You have pretty much a maximum of four or five hours of bottom time in Alvin, and then the transit up can take two to four hours.

The navigation is still carried on at the mother ship?

We know where Alvin is at all times. We have a sonar navigation net that we deploy.

What's the name of your oceanographic vessel?

The vessel that Alvin is currently on is the Atlantis II. We have three ships in Woods Hole that we operate: the Knorr, the Atlantis II, and the Oceanis. The Knorr is queen of our fleet; she's 235 feet long, about 3,000 tons, and has the advantage of an omnidirectional propulsion system, which was one of the key factors in our finding the Titanic. You have two joysticks: one for the forward thrusters, with a cycloidal propeller, and one for the rear cycloidal propeller. You just push the joystick in whatever direction you want to thrust, and the ship does the rest.

Left: the bow of the Titanic from above. Note the two capstans and the port and starboard chains. © Woods Hole Oceanographic Institution.

Below left: Jason Jr. exploring the Titanic. Below right: The prow of the ship. © Woods Hole Oceanographic Institution.

Back to the Pacific.

Well, the East Pacific Rise is a bit north of the Galapagos. Our mission was to map as large a part of that ridge as we could in the allotted time of three weeks. It took us about a week to get to the site, so we had something less than 14 days. That voyage was

sponsored by the National Science Foundation; in fact, they block-fund their oceanographic ship time, so that three weeks was National Science Foundation time allotted to us.

Other scientists have time allotted to use a cyclotron, a synchrotron, or some other type of large machine; you have time allotted to be in the ocean?

Exactly. In fact, ships are expensive resources. The big ships that we use for ocean research cost anywhere from $8,000 to $18,000 per day. That includes a full complement of crew, food, and provisions. In addition you have to pay the salaries of everyone on the mission. This Pacific expedition was Argo's first scientific mission—it was Argo that found the Titanic. Argo is a towed vehicle that is essentially a video platform, and it also has sonar capability. The major advantage of Argo over something like Alvin is the amount of bottom time you have. We were able to keep Argo down for as much as 72 hours, just mapping constantly, much to the jubilation of the geologists. Personally, I've never seen so many rocks in my life.

In what capacity did you act?

I was there primarily as an observer. I would say that my mission, along with Bob Ballard, was to recommend future directions in ocean imaging. It was important to understand the state of the art and what the consumers—geologists, in this case—were interested in. One of the other things we were looking at was a mini command control center for Argo, where we could display all of the different kinds of multisensor information that we collected in a way that was more easily interpretable by the scientific staff members who were making decisions; in other words, we were collecting video information, sonar information, and navigation information.

Were there marine biologists aboard too?

That's correct. We had a woman who had done a lot of vent biology, mapping the distribution of vent animals, which was her main interest—the ecology of these thermal vents.

You found that the ecology was so different that you have different life forms there?

They don't take aerobics classes, they take anaerobics classes. They also seem to be able to tolerate large temperatures, which is interesting from a biochemical and biophysical point of view, how organisms have evolved to tolerate large temperatures. There are some claims in the literature that these organisms live at temperatures hotter than any known proteins can tolerate.

That could be the basis of science fiction, that type of life form.

It very well could be. In some sense these life forms may be more primitive. One of the interesting questions that needs to be answered is why these isolated vents all show the same life forms.

You described Argo as having video equipment at one end and lights at the other. Can you explain the principle and how it worked out?

Right, that came out of my research, so I'm glad you asked. I think you have to look at underwater imaging as a system with an input and an output. As an engineer you want to optimize the throughput to that system. So you start at one end, and you ask yourself, "Can I do any better?" and you work your way through to the other end, asking yourself the same question many times. For us, the lighting configuration was really our front end, in other words, the part of the system where we're collecting images. There were some concepts around in underwater lighting in the mid-70s that advocated displacing the camera and the light in a vertical direction, that is, hanging the light over the camera. In the underwater imaging regimes that we're operating in, we're primarily backscatter-limited. I've described that as driving through a fog bank; everyone knows that turning on your brights doesn't help. Well, it's unfortunate that people have been turning their brights on in the ocean for quite a long time and not realizing it, because more power won't help at all. What we did was model the underwater imaging process in our computer, using the known physics. We decided that there were large gains to be made by displacing the camera and the lights at opposite ends of the sled.

Same level?

In Argo, that's correct. We decided that the horizontal displacement had a great deal of advantage over the vertical. It has to do with the amount of volume that you intersect for backscattering.

That came out of your modeling?

Correct. Those results were presented at the Ocean Optics VII session today.

You have a third vehicle. Would you mind describing that?

Yes, the Angus. Angus has been the workhorse for the oceanographic community for quite some time now, something on the order of 10 years I imagine. The basic idea of Angus is that it's a 35 mm camera sled. Unlike Argo, though, Angus has no video on it, and there's no video telemetry system. With the Argo you can

In the underwater imaging regimes that we're operating in, we're primarily backscatter-limited. I've described that as driving through a fog bank; everyone knows that turning on your brights doesn't help. Well, it's unfortunate that people have been turning their brights on in the ocean for quite a long time and not realizing it, because more power won't help at all.

actually see what you're looking at, but in Angus, you are more or less flying blind, clicking away your 35 mm cameras. The great advantage is resolution, because currently our video telemetry link for Argo does not have the resolution of 35 mm, nor do our video cameras, for that matter. So the pictures that most people have seen of the Titanic, the higher resolution ones, are in fact Angus pictures that were taken with 35 mm.

So you find the sites you want to record and then send in Angus?

Right. In the Titanic mission, we flew Argo, we mapped the site, and then we went in with Angus for a high resolution survey. On our most recent expedition to the eastern Pacific, we mounted a 35 mm camera on Argo itself, and we were able to snap our 35 mm pictures and know exactly what we were looking at. So Angus was never used in that mission, although it was on the ship all the time in case there was ever any kind of problem with Argo. It looks pretty much like Angus will now be retired, and Argo will be a multifunctional instrumentation platform for video, 35 mm, and sonar as well.

Do you have other vehicles in the planning stage?

Yes. The next generation of ocean-going submersible vehicles is actually going to be robots, because Dr. Ballard secured funds for a two-part program called the Argo/Jason program. The first part of that program is reasonably complete now, the Argo part, and we're forging ahead in the robotics part, which is Jason. Jason, in fact, is a robot that will be the son of Argo.

Are these acronyms, or are they based on the Quest for the Golden Fleece?

They're based on the Golden Fleece. Jason and the Argonauts, that's correct. Jason will be developed within the next two or three years. In fact, I think we have some field trials coming up for Jason this summer. The basic idea is that of a hierarchy of magnifications. In other words, Bob Ballard has said on many occasions that 99.99% of the ocean is extremely boring, but it's really hard to find the other 0.01%. So the basic idea is that we're going to have a hierarchy of views; the lowest resolution view will be a sonar view. There are now a number of sonar platforms that get resolution on the order of 100 meters or so with picture sizes on the order of several kilometers. The next generation of imaging platforms, of which Argo is an example, is composed of vehicles or platforms whose resolution is on the order of meter or fractions of a meter. The third instrument in this chain is Jason. Being a robot, Jason can obviously get extremely close, and so the idea is that you sort of telescope in on the interesting features through this hierarchy of imaging platforms. Jason will be operational within two years and will be housed by Argo, so that when Argo finds something interesting, Jason can actually go down and explore it. Jason will have manipulators to pick things up and put things down. One thing we haven't talked about is the idea of telepresence, which is what the lab is founded upon. The idea is to telemeter oneself to another location through other sensors. And in our case, Jason is envisioned to be somewhat of an extension of the human. What it sees you will see; what it feels you will feel. There are some very interesting interface ideas in robotics that are also being investigated in the lab.

Let's go to the Titanic then. How does a project like this come about? In your talk you mentioned the Titanic cult aspect to describe the amount of interest in the folk history of the Titanic.

I don't know. It's a good question in contemporary sociology. I can speculate, although my speculation is certainly no different from anyone else's. I've been asking a lot of people who are fascinated with the Titanic the very same question. One of the answers I got, for a certain generation of individuals, was that commercial aircraft didn't really exist when they were growing up, and the spectacular disasters of that era were, in fact, oceanographic disasters. I also think this vision of all people ultimately being equal, whether you're John Jacob Astor and you're the richest man in New York or you're down in the steerage of the ship— before the forces of nature, which man cannot control—is a vision that appeals to humanity.

How did the deep submergence people get interested? Did somebody come to them?

No, it was Dr. Ballard's idea to mount an expedition to the Titanic. I think that Dr. Ballard would like to be remembered as one of the foremost underwater explorers in history. If he had been born 150 years ago, he'd have been there with Lewis and Clark. Had it been 250 years ago, he would have crossed the Atlantic on a square rigger. He's just that kind of individual.

Was this Titanic expedition a joint operation with French oceanographers?

Right. Dr. Ballard has had a long history working with the French. There was a geological mission called the FAMOUS Mission in the 70s to explore part of the mid-Atlantic ridge. We've enjoyed a good relationship with the French for many, many years. Our ultimate interests are to go to the Mediterranean to do underwater archeology, and we are working hard at becoming friends with countries that border the Mediterranean—partly in the hopes of pursuing those interests in the future. Also, I might add, the French have some of the most sophisticated marine technology in the world, including several diving submersibles, and they continue to be major leaders in underwater vehicle technology.

Their crew was assigned to do a sonar map; how did you know where to start looking? How did you ever pick the haystack?

Well, there are quite a number of historical documents from that time because the event was so widely publicized. I think that Dr. Ballard and his coworkers, who, incidentally, included the Titanic Historical Society—the world's experts in Titanic history—had, in fact, gleaned from the many reports about the location of the ship, a reasonable—10 mile by 10 mile—area in which to look. In other words, there were several reports of sightings of the ship or of where the passengers were picked up, and a lot of those agreed with each other; some didn't, but the differences were on the order of 10 miles or so. So we were fairly certain, because there seemed to be some consensus among the rescue ships, that there was a reasonable area in which to look.

They knew where they picked up the lifeboats?

Exactly. And also debris from the ship itself.

And you took about 100 square miles?

Correct. There was about a 10 mile by 10 mile area—100 square miles—in which we were fairly confident that the Titanic would lie. Now in the analysis of previous missions—in other words, you might ask yourself why previous missions had failed to find the Titanic—our analysis essentially led us to the conclusion that they simply did not have enough time on target. And, in fact, one area of Dr. Ballard's expertise is search procedures. There is, in fact, an engineering science of search procedures, and it seemed to us that some of the people who had been in those earlier expeditions were unaware of this science.

Were some of them private expeditions?

That's correct.

Hoping to get booty or a claim?

I think the Titanic really is so deep that the major benefit of finding the Titanic would, in fact, be just that—finding it.

And there have been people searching?

Yeah. There's a Texas millionaire, Grimm, who sponsored at

least two expeditions to the site, in fact.

Was his in the same vicinity of your 100 square miles?

Yes, they were looking in the right place.

They were in the haystack, eh?

Yeah, they just couldn't find the needle. And we know why but it's not something that we're interested in publicizing.

O.K., so there were some private expeditions.

That's correct. Actually, they benefited ocean technology because a fair amount of money was pumped into the development of new underwater technology—on the order of several million dollars—which resulted in some new sonar technology being developed. So actually, the searches were quite beneficial even though they failed. And there's a big controversy now. Grimm claims that he, in fact, did see the propeller of the ship, which we looked at; it turned out to be a rock. And he's laying some sort of claim to the site. There are lots of crazy rumors going around about an old submersible called the Alluminaut, which may be refitted. And we've even heard some talk about people giving tours to the site, actually taking people down in the Alluminaut to see the Titanic. So, in fact, the Titanic did lie within our 10-by-10 mile square. Now you may have been alluding to the French-American plan: the French were to do a sonar site survey of the entire site. They were bringing along a very nice instrument that had been developed by Thompson CSF in France—the French do make some of the best sonars in the world. Unfortunately, the French were unable to map the entire area because of the strong surface currents. Their constraints were financial, as were ours. Within their allotted time, they were able to map only 80 percent of the site. So when we got to the site, it was our job to explore the remaining 20 percent, or as much of it as we could with Argo. Now one of sad parts of the story is that they were very close to the Titanic, but not close enough to see it, really.

You got a quarter of a way through your allotted time at sea?

Actually, it was quite close for us, time-wise, because we knew we had a three-day transit back to Woods Hole, and we found the ship about two or three days before we had to start back.

You also indicated that Argo sort of stumbled on to it.

Well, we knew that the debris field was something like a mile wide. So you may have been thinking of Alvin stumbling upon the Galapagos. Argo was part of a well-planned search for the Titanic. We knew pretty much what the debris field was going to be, and our search pattern was oriented around searching on a grid that was appropriate for that.

What did Argo see?

Well, the first thing that was seen, in fact, was one of the large boilers from the Titanic. They were something on the order of 25 feet in diameter.

Was the video good enough so that there was no doubt?

Oh, yes. It was just amazing. The video's certainly that good. In fact, they discovered this large circular object and then immediately went to one of a number of books about the ship and looked it up. And it was unmistakably one of the boilers from the Titanic. And that, in fact, I think, is one of the morals of the story in a sense. With sonar identification of targets, you have a lot more ambiguity in your identification, whereas we are visually trained, being visual beasts—the largest part of our brains is, in fact, dedicated toward vision. And that kind of information, perceptually, is very easy to assimilate. So visual imaging does have quite an advantage over sonar, and the finding of the Titanic is, in fact, testimony to that. It was unmistakable that that was a boiler from the Titanic. Now once that was found, they knew that the Titanic itself, or herself, would be within a mile or so. And so, they pulled Argo up a couple of hundred meters so that she would be clear of the bottom, and then actually transited the ship around to see if they could find any large excursions in the bottom and, as I mentioned last night at the lecture, the hulk of the Titanic was found with the most archaic of

We are visually trained, being visual beasts—the largest part of our brains is, in fact, dedicated toward vision. And that kind of information, perceptually, is very easy to assimilate. So visual imaging does have quite an advantage over sonar, and the finding of the Titanic is testimony to that.

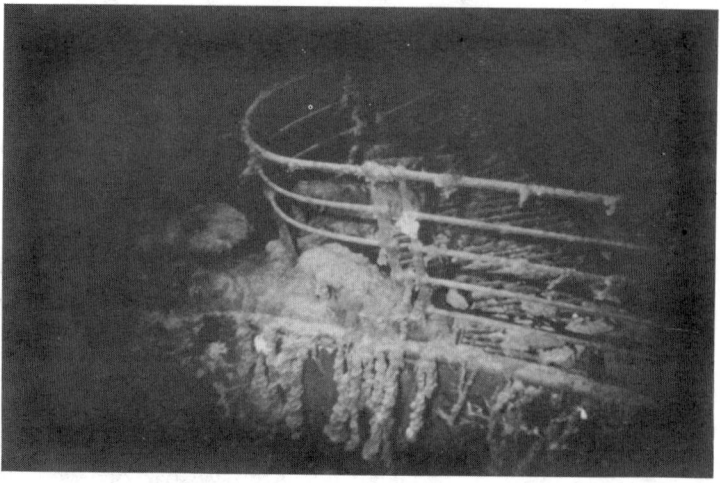

Railing of the Titanic.
© Woods Hole
Oceanographic
Institution.

all underwater exploration devices—that is, a simple bottom pinger—because they were able to notice a hundred-foot excursion in the bottom over a distance that seemed like it would be compatible with the hull of the ship. It was one of the curiosities of the story. So we did a site survey as best we could with Argo, the video. And then we pulled that up and lowered Angus, the 35 mm camera sled and did a close-up survey of the Titanic. I'd also like to point out the hazard of flying a platform—almost blind— close to a large object from which you suspect there would be wires hanging all over the place. In fact, I did mention last night that we banged into one of the smokestacks on the ship at one point, and got some Titanic paint on Argo. And so within the remaining day or two, we worked very intensely to get as much information as we could within that amount of time. And the results have been shown in the December (1986) issue of *National Geographic*. Some of that footage was also shown on the Disney Epcot special.

What things of significance came out of this expedition for advancing the field of image processing? We've already mentioned the underwater illumination.

Well, Argo itself is integrally related to image processing. The system as currently configured can be operated in two modes: continuous video or snapshot video. In the continuous video case we are limited to doing our processing on the fly, as we say in the industry. All of the images we observe are processed through our image processing equipment. We can, in this case, manipulate the gray levels of the image via look-up tables. We can also do frame averaging in order to temporally filter out some of the SIT (silicon intensifier target) camera noise.

You mentioned that there were two imaging modes.

That's right. In our snapshot mode we use a set of strobe lights instead of our continuous illumination lights. In this case we can actually take a frozen frame and do some more advanced processing with it. This is still somewhat experimental and research-oriented. I am currently developing algorithms to enhance not only the gray level information but also the resolution of these images. These techniques are somewhat similar to conventional areas of image processing but with some interesting new twists due to the uniqueness of the application.

Are the French going to follow up on the Titanic?

Yes, that's a good question, actually. The French, to our knowledge, are planning to go back, and we recently had some meetings with them. They're planning on taking their submersible to the Titanic site—a deep submergence submersible called Nautile. And we've recently had some talk with them about coordinating rescue operations between the two submersibles in case there are any problems, so we'd like to keep the spirit of international cooperation.

How far down was the Titanic?

Fourteen thousand feet!

Heinrich Robrer

Scanning Tunneling Microscopy

Gerd Binnig and Heinrich Robrer. (Photo and all illustrations in this article courtesy of IBM.)

A Microscope That Makes Atoms Visible

Have you ever heard of the scanning tunneling microscope? It's a microscope invented by Gerd Binnig and Heinrich Rohrer of IBM's Zurich Research Laboratory. They were awarded the Physics Nobel Prize in 1986 for that novel instrument—an instrument that makes it possible to see individual atoms on a substance's surface. And you can discriminate differences of 1/100th the size of the atom.

This scanning tunneling microscope is a further development in microscopy whose history goes back to the 15th century. The optical microscope made clear the existence of bacteria and pathogenic organisms, and provided detailed knowledge of the cellular structure of organisms. But because the wavelength of the light used by them is about 2,000 times larger than the size of atoms, it is impossible to use optical microscopes to see atoms.

At the beginning of the 20th century, x-rays and electron waves, whose wavelengths are shorter than those of light, began to be used. But with x-rays, the condition of only a certain configuration of a substance's atoms can be observed. With electron micro-

This article is based on an interview with Heinrich Rohrer of IBM's Zurich Research Laboratory, first published in *Chuokoron*, Tokyo, April 1988, and revised by Heinrich Rohrer. Used by permission.

scopes, individual atoms can be observed, but only when they are arranged in a certain sequence or configuration.

In contrast, the scanning tunneling microscope permits one to see individual atoms regardless of their sequence or configuration. When talking about "seeing" atoms, it is probably more appropriate to say that the electrons of the atom are "felt." An atom is comprised of a small nucleus at its center and electrons that envelop the nucleus like a cloud. With the scanning tunneling microscope, an atom can be observed by "feeling" or "touching" these electrons.

By "feeling" the atoms in this manner, one can tell how strongly an atom is bonded to neighboring atoms, and how hard each atom is. It's rather like being able to feel whether someone has a fever or whether his skin is dry by touching his face. "Actually, temperature sensors derived from the scanning tunneling microscope have been recently developed," Rohrer says. "They can feel temperatures of very small spots, very much smaller than what can be seen by an optical microscope. Likewise we have learned how to 'feel' spots of liquid, rather than just solid surfaces."

The scanning tunneling microscope touches a substance at the atomic level and allows one to derive all kinds of concepts about the substance. However, only a substance's surface properties can be observed.

A Surface is the Invention of the Devil

Since the surface is the first thing to meet the eye, naturally one would think that it is surfaces that are understood in greatest detail. But it is surfaces that theoreticians abhor the most. The famous theoretical physicist Wolfgang Pauli was alleged to have said that "surfaces were invented by the devil." Because the symmetry observable inside an object falls apart on the object's surface (that is, in contrast to internal atoms being surrounded in all directions by other atoms, only the lower half of surface atoms are surrounded by others), problems relating to surfaces arise that are often too complex for theoretical calculations.

New detailed knowledge about such surfaces obtained by using the new microscope will help science and technology go on to make great advances. For example, now that computer elements are getting smaller and smaller—the smaller they get the greater the surface to volume ratio becomes—the contribution of surface area in relation to the element's function grows ever more crucial. In the chemical industry, new catalyzers for conducting chemical reactions more efficiently are always being sought, so knowing in detail the electron structure of a catalyzer's surface will greatly facilitate the discovery of new catalyzers.

Because many important reactions in organisms take place through membranes, some important discoveries (in the organic sciences) are probably in the offing. This is especially so because the scanning tunneling microscope permits the observation of membranes while they still contain water, i.e., in a life environment. This is in contrast to the electron microscope, in which case one would have to be satisfied with staring at surume (dried squid) when it was ika (live squid) that one wanted to see.

"Scanning tunneling microscopy has opened many new ways to deal with individual atoms and molecules, and will play a very important role in our drive to nanometer-scale science and technology. This will pose a tremendous challenge for science and technology as well as for society at large to deal with even more complex issues than were introduced by microelectronics or microtechnology," Rohrer says.

Birth of the Scanning Tunneling Microscope

In the 70s, IBM's Zurich Research Laboratory investigated Josephson junctions as possible new elements for computers. Rohrer recalls, "The tunneling current through extremely thin oxide layers was a crucial but also problematic element in this project. It was also becoming an important issue in many other technological applications. It, therefore, appeared important to understand and test the electronic properties of such thin insulating layers on a scale very much smaller than had ever been tried before."

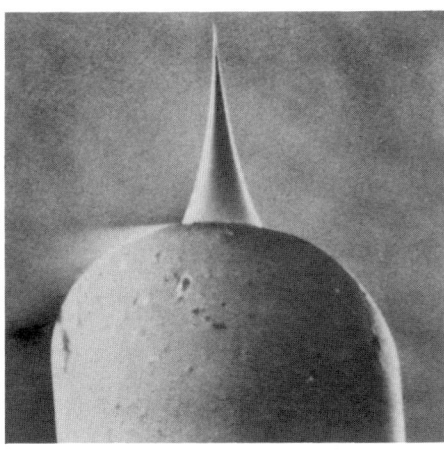

The tip of the STM.

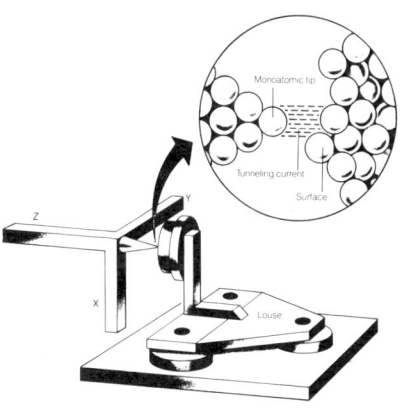

The scanning tunneling microscope.

Since it appeared also to be an interesting scientific problem, Binnig and Rohrer set out to tackle it. The tunneling current was a natural tool for their investigation. When the tip of a probe is brought close to the surface of a substance, as if applying a brush very lightly on a cloth, the electrons at the tip of the probe almost come into contact with the electrons on the surface of the substance; then an electric current known as a tunnel current flows. "Feeling" and "touching" means that the tip of the probe comes in touch with the surface of the substance through this tunnel current. It will start flowing when a probe is brought to within a few angstroms of the surface of a sample. The quantity of the current is so sensitive to the distance between the probe and the surface that it will change by as much as 1,000 times when there is a change in distance equivalent to the diameter of one atom. "By making use of the tunnel current's exceptional sensitivity to the change in distance," Rohrer recalls, "we figured it would be possible to measure with extreme precision the vertical positioning of the atoms lined up on the surface of the sample. If we could get accurate feedback of the change in the current, we would be able to scan the surface of the sample by keeping the tip of the probe at a constant distance from the irregularities on the surface. Then we could record the movement of the probe and thus obtain an image of the surface atoms either on the CRT screen or on the plotter."

Immediately after starting to make a scanning tunneling microscope, however, Binning, Rohrer, and Christoph Gerber, who joined them in this venture, were confronted with a multitude of technical problems. For example, how was the probe to be moved on the infinitesimal scale of atoms? How could they get rid of the minuscule mechanical vibration of the probe as it traces the distance from the surface of the sample? What was the best shape of the probe and how could it be made?

It took 27 months before they solved all those problems and created a microscope. It was on March 16, 1981, when they could control the tip sufficiently to measure the tunnel current. Some months later, they obtained the first image where single atomic steps on a metallic surface could be resolved.

Do Not Fear Failure

Binnig and Rohrer had worked before in the field of superconductivity, so they had some experience with tunneling phenomena. Moreover, Rohrer had previously measured the change in length of superconductors, at the normal-superconducting transi-

> *Management, at all levels of the laboratory, [should] consider as its primary task the support and encouragement of its scientists rather than taking upon itself the generation of ideas.*

tion, to the accuracy of one angstrom, and wrote a thesis on it for his PhD. So he felt it could be done. So did Binnig. Still, as they knew almost nothing about microscopes and surface science, the probability for success did not seem very high to others. Rohrer recalls, "Most our colleagues said, 'That will not work.' In a way we were fortunate, because we undertook the project lightly. We were not afraid of failing. Anyway, we made no excuses, such as we would not succeed unless we did this or that first, or unless we knew about this or that beforehand. We just had this feeling that we could do it, and did everything we could to make it come true."

"The advice I can give now to someone undertaking a new project is: Do not be discouraged by the many difficulties that may stand in your way, but just commit yourself and jump over the barriers. If you keep on doing only the things that can be done in the realm of general knowledge, you will never achieve anything new."

What Is IBM's Zurich Research Laboratory Like?

It is well known that after Binnig and Rohrer received the Nobel Prize in 1986, their colleagues, K. Alex Muller and J. Georg Bednorz, also received the Nobel Prize, in 1987, for high-temperature superconductivity. Because of that, the question often arises "What is IBM's Zurich Research laboratory like, in that it has produced Nobel laureates two years in a row?" Rohrer answers by saying that the Nobel Prize is given to persons who have succeeded in a truly innovative project and not to a laboratory. A laboratory cannot produce Nobel Prize winning researchers simply because it is determined to do so. However, a laboratory can attract very good researchers and provide a stimulating environment, such that one of them might win a Nobel Prize.

"Our laboratory provides us with a very favorable environment for research. That is to say, first, the laboratory offers researchers the facilities and means that are necessary for them to conduct their research freely and in the direction they desire," says Rohrer. He goes on to say that the laboratory provides them with an environment in which they can devote themselves to studies without exposing themselves to too much pressure from the outside, meaning that they do not have to constantly promise to solve one problem after another. It seems that promises arouse expectations, and expectations bring on more and more promises. Of course, scientists have to live with expectations of those who support them, but it is important that the expectations are in harmony with what the scientist can do and is doing.

Science is a quest for new ways of thinking and new perspectives.

"What is important is the atmosphere in which researchers can devote everything to study, without bothering about anything else," Rohrer states. "For example, I heard that Georg Bednorz and Alex Muller had never told anybody about what kind of research they were engaged in. They were able to pursue their research for two years, freely and without any interference from others. So the best course of action for management of a laboratory is to recruit researchers of outstanding qualification and then provide them with an environment in which they can engage in free research."

Rohrer feels strongly that management, at all levels of the laboratory, consider as its primary task the support and encouragement of its scientists rather than taking upon itself the generation of ideas.

IBM's Zurich Research Laboratory has about 200 members in all, of which the Physics Division, to which Rohrer belongs, has about 40. There are other departments engaged in diverse projects, such as communications, computer science, and laser science and technology. At the laboratory, there is an intellectual atmosphere that encourages different points of view and new ways of doing things. And there are capable technicians who are experts in their respective fields. Binnig and Rohrer's research of the scanning tunneling microscope started in that favorable environment.

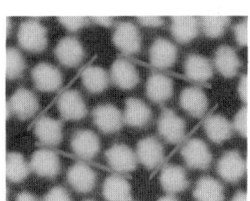

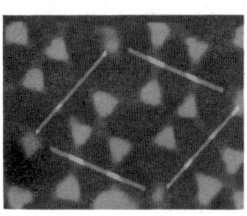

"Other people certainly have other ideas of how things should be done and of how things happen. Perhaps, they should also contact the Institute of Immunology of Hoffmann-La Roche in Basel, Switzerland, a pharmaceutical company, in which some 60 researchers likewise enjoy a great deal of freedom in their research. They also recently produced two Nobel Prizes," Rohrer says.

Binnig and Rohrer are often quoted as saying that their research, like that of Georg Bednorz and Alex Muller, did not

Above: Two STM images of the silicon (111)-(7x7) surface with the 7x7 unit cell outlined. The (111) refers to a diagonal plane cutting through the 3D trapezoidal structure of the unit cell. The 7x7 unit cell of the surface simply means that the areal projection along the (111) plane is 49 times larger than that of a unit cell in the bulk structure of silicon. Top: The overall surface topography showing the 12 adatoms, or atoms at the topmost layer of the surface. The vertices of the trapezoid end in corner holes (that is, "windows" to the layer beneath). Bottom: This shows the surface states on the atomic layer beneath the adatoms.

At right: A schematic side view of the silicon (111)-(7x7) surface showing three kinds of surface states. The six topmost lobes are the states on the adatoms. The four middle lobes are states on the atoms below the adatoms, and the remaining two lobes are localized in the deep corner holes of the 7x7 unit.

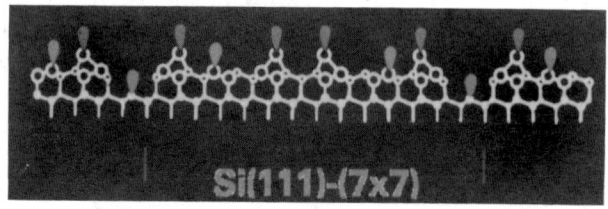

require much money. While this is true for the specific projects as such, Rohrer feels it is inappropriate to separate the two projects from the total operation of a laboratory that provided the underlying groundwork for basic research in many respects. Such an operation can be quite expensive. It is a symbiotic relationship. The scientists cannot exist without good support and, of course, a laboratory is nothing without ideas.

Searching for New Perspectives

Scientists, when setting out to do something new, should not premeditate on its possible applications or commercialization. Setting a practical goal narrows your scope of thought. If the study is on something genuinely new, Rohrer believes it will always be the source of significant applications to come.

"Science is a quest for new ways of thinking and new perspectives," Rohrer says. "Thinking of science in terms of 'good' and 'bad' or 'useful' and 'useless' at the outset is restrictive. Science may be applied for 'good' or 'bad' use later on, but essentially science is free of values. I am not saying, however, that scientific inquiry should be performed in limbo, disconnected from social reality and responsibility."

Moreover, Rohrer believes that scientists should not think in terms of whether the research is superior or inferior to that of others. If what you are doing is genuinely new, you cannot compare it with other studies. "Go with your perceptions, your feelings," he says. "We are, after all, in the business of pushing out the frontiers of knowledge."

Frederick Su

Chaos

*Without dimension, where length, breadth, and height,
And time and place are lost; where eldest night
And chaos, ancestors of nature, hold
Eternal anarchy ...*

Paradise Lost *(1658-1665),*
Book 2, lines 893-896
John Milton

Man's predilection for simplicity and Nature's innate complexity provide grist for the scientific mill. The turbulence of a rushing stream over rocks, the rising swirl of cigarette smoke, weather patterns, the roll of the dice, epileptic seizures, heart arrhythmias (Figure 1), laser instabilities—all of these phenomena had been previously perceived as totally random events, chaotic in the sense of Milton's lines. These ubiquitous ambiguities and anomalies of nature were formerly unpredictable, unwelcome, and quietly swept under the rug as untenable in the domain of pure science. Rarely were closed-form solutions found to the nonlinear equations that modeled such phenomena. Termed intractable, they were only made less so with the advent of the great number-crunching computers.

But now, utilizing the graphics capabilities of computers that can handle large data sets, researchers are achieving a visual "feel" for the solution. This computerization of mathematics has given rise to the study of chaos. "In the last 15-20 years, the whole development of nonlinear dynamics, i.e., the study of nonlinear phenomena, has changed the meaning of chaos. Now we use the term 'chaos' to refer to a series of events that look random but, in fact, point to a deterministic development from nonlinear equations," Robert Gioggia of Widener University said. And Neal Abraham of Bryn Mawr College went on to explain, "Chaos is midway between the simplest forms of nonlinear dynamics and the most complex, which is turbulence. That is, chaos is deterministic irregularity." And the explosion of interest in studying the newly

Reprinted from *OE Reports*, March 1990. Frederick Su is a technical consultant for SPIE. He received a PhD in theoretical physics from the University of Connecticut in 1979.

defined chaos arose from the measure of predictability afforded by the strange, or chaotic, attractor.

What is chaos? What is a fixed point attractor, limit cycle, strange attractor? What about period doubling, Lyapunov constant, Feigenbaum constants, Lorenz butterfly, Birkhoff bagel, and Rossler band? What does iteration of a mathematical function have to do with some turbulent phenomena? What does chaos have to do with fractals?

Why all the excitement? Because, with chaos theory, researchers have another tool with which to predict the behavior of nonlinear phenomena. There is a measure of determinism or predictability in what were seemingly random events. And, in the

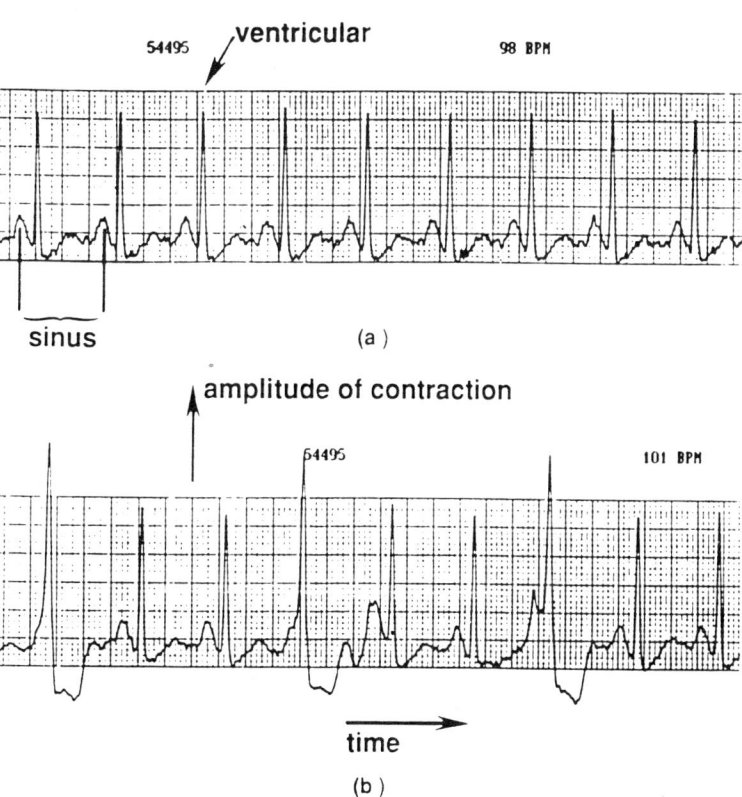

Figure 1. The normal heartbeat is determined by two sets of nerve bundles within the heart, the SA (sinoatrial) and AV (atrial ventricular) nodes. In a normal heartbeat (a), they work together to give a regular pattern. In an arrhythmic heart (b), the amplitudes of both the ventricular and sinus beats are irregular, something like a car engine missing. This is dangerous since it leads to lower blood pressure. [10] Chaos is useful in modeling the nonlinear dynamical heart. New research points toward a healthy heart being "chaotic" and aged and diseased hearts as exhibiting regular periodicity. [14]

LORENZ EQUATIONS

For the convection of heat in a fluid layer:[*]

$$\frac{dx}{dt} = -ox + oy$$

$$\frac{dy}{dt} = -xz + rx - y$$

$$\frac{dz}{dt} = -xy - bz$$

where: x = the intensity of the convective motion
     y = the temperature difference between ascending and descending currents
     z = the distortion of the vertical temperature profile from linearity
     o = the Prandtl number of the fluid
     r = the Rayleigh number of the fluid to be varied by changing the size of the sample.

Table I

midst of this, there suddenly appeared some universal constants, first derived from pure mathematics and then found in physical systems. Not surprisingly, this piqued the interest of many researchers around the world.

Chaotic attraction

The basis for this was laid in 1963 by Edward Lorenz of the Massachusetts Institute of Technology. Trying to understand why weather patterns were so unpredictable, he first modeled the atmosphere of the planet as a fluid using the Navier-Stokes equation, and with simplifying assumptions, ended up with three first-order coupled differential equations (Table 1). Using a digital computer to run his model, he plotted his solutions in a 3-dimensional phase space. He found that small differences in initial conditions led to exponential divergence of the solution as time passed. This "sensitive dependence on initial conditions," a phrase attributed to the great French mathematician, Poincaré, himself, showed up on Lorenz's computer screen as a butterfly or mask (Figure 2). Behavior of a trajectory on its surface was unpredictable, but there was one general constraint. No trajectory, for the most part, could leave its surface. This was the first original strange, or chaotic, attractor that was discovered.

What did this mean? It meant that in terms of weather, there were certain constraints. It would be very unlikely for Boston to have a 200 degree day in December or Los Angeles to have a -30 degree day in July. Such events were not on the surface of the strange attractor. But a person could only model the weather in the short term, because the initial conditions from one day to the next, even though minutely different, led to exponential divergence of the solutions with time. Thus, long term predictions of the weather could not be done.

Let us try to better understand what is meant by an attractor. A pendulum swinging through small angles in a frictionless environment is an example of linear periodic motion. We can visualize the pendulum in a two-dimensional phase space (also called state space) of velocity and position (Figure 3). The motion of the frictionless pendulum describes a closed ellipse. However, once friction is introduced into the problem, the motion of the bob spirals down to a point in phase space, i.e., the trajectory comes to rest at a fixed point. "It is said to be a 'stable fixed point' because, if the system is infinitesimally perturbed from this value, the perturbation disappears asymptotically," Abraham said. And since the system attracts all nearby trajectories, it is an attractor. Thus, the rough definition of an attractor is a state to which a dissipative system settles down, where a dissipative system is one whose energy is lost due to friction or some other form of damping. Thus, the solution where the pendulum hangs motionless is called a fixed-point attractor; it has zero dimension (a point has no dimension), no frequency, no period. [3]

The next most complicated attractor is called the limit cycle. An example of this is the behavior of the metronome. It can be our simple pendulum again, but this time made up of a bob on a rigid

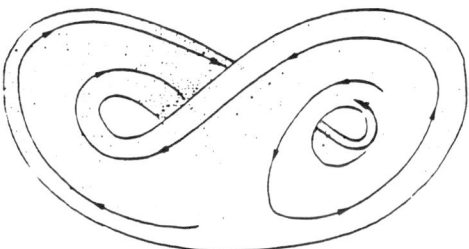

Figure 2. The Lorenz butterfly or mask depicts the trajectories that form a strange attractor based on solutions from a simple model of atmospheric convection. [3]

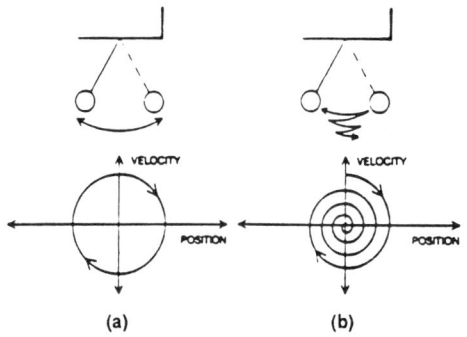

(a)　　　　　　(b)

Figure 3. (a) The phase space diagram of an ideal pendulum is an ellipse. (b) The phase space diagram of a real pendulum (one damped by friction) shows that the oscillation dies down to zero, i.e., its motion is finally attracted to a point. [2] (From "Chaos," by James P. Crutchfield, J. Doyne Farmer, and Norman H. Packard. Copyright © 1986 by Scientific American, Inc. All rights reserved.)

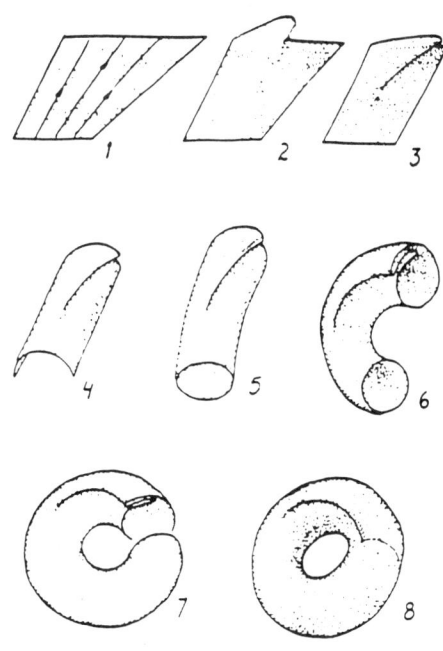

Figure 4. The Birkhoff bagel provides the superstructure on which the strange attractor resides. This strange attractor covers a folded surface where trajectories that are initially close together diverge exponentially. The fold also brings trajectories that were once far apart close together since the upper and lower levels of the fold are close. This provides the reinjection or molding that is another key feature of chaos. [3]

Reprinted with permission from *Scientific American*, December 1986, page 53.

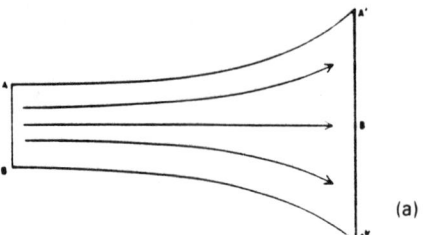

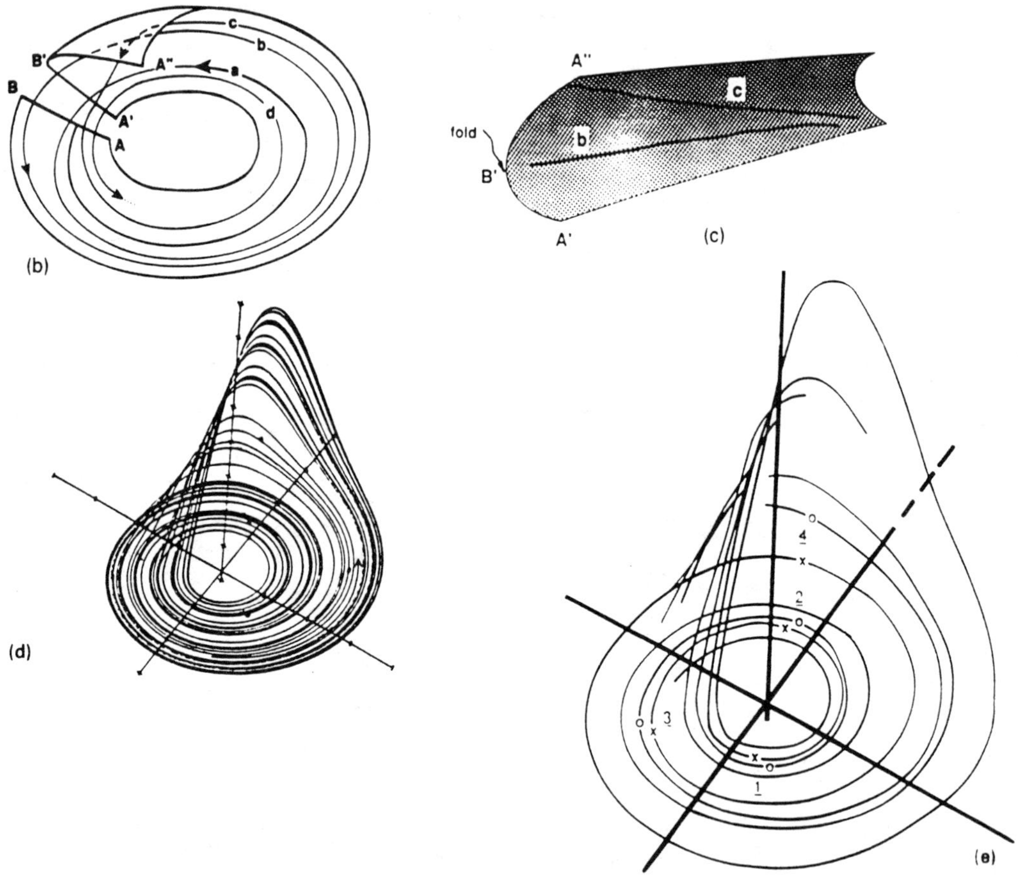

Figure 5. The Rossler band is the simplest strange attractor. (a) We start by stretching a surface as shown. (b) It is then stretched into a circle and folded. Note that orbits a, b, c, and d do not intersect. In (c), it is clearly seen that the fold allows orbit c to cross over orbit b without intersecting it. With a strange attractor, paths in phase space can never intersect because their futures are governed by where they are and, if two paths happen to meet, they then have the same future and the solution is periodic, contradicting the aperiodicity of chaos. (d) The surface is "joined" to form the Rossler band. Exponential divergence occurs only for nearby trajectories and only in the short term. Eventually the trajectories return to be near each other since the attractor is bounded and, in this phase, they converge to each other and then diverge again. (e) This less cluttered redrawing of (d) easily shows the exponential divergence, at different stages 1, 2, 3, 4, of an X and O that were initially close together on nearby trajectories. The constant in the exponent defining how fast the trajectories diverge in time is called the positive Lyapunov constant. [2, 6] (Fig. 5a from "Chaos," by James P. Crutchfield, J. Doyne Farmer, and Norman H. Packard. Copyright © 1986 by **Scientific American,** ***Inc. All rights reserved. Fig. 5d courtesy of James P. Crutchfield)***

rod and driven periodically by mechanical or electrical means. The system can be slapped and pushed around, but as long as it is not destroyed, it will always return to periodic motion of the same frequency and amplitude. Another example (in this case an approximate illustration) of a limit cycle is a normal heart. While its frequency may increase due to fright or exercise, it always settles down to a resting rate. In each case, the periodic solution is an attractor, since all motion is attracted to it. Limit cycles are closed orbits in phase space.

Two simple examples of phase space trajectories of chaotic attractors are shown by the Birkhoff bagel (Figure 4) and the Rossler band (Figure 5). Both structures demonstrate: (1) exponential divergence of neighboring trajectories (how fast they diverge depends on the Lyapunov constant), (2) bounded trajectories that are reinjected into the neighborhood of places they left, and (3) mixing or folding of trajectories. As a corollary, any one path cannot intersect itself at any future time, because to do so would imply a periodic orbit no matter how much time has passed. And chaotic orbits are not periodic. On both the Birkhoff bagel and Rossler band, these requirements are accomplished by first stretching the sheet at one end and then performing the fold. As can be seen in Figures 5b and c, that extra dimensionality afforded by the fold allows a trajectory to cross over or under another trajectory (or previous path) without actually intersecting it.

Universality

The universality defining chaos has its roots in pure mathematics. It was Mitchell Feigenbaum, then at Los Alamos, who in the mid-1970s set the stage for much of the excitement today. If we take a simple function such as a parabola defined by the equation $f(x)= y = 4\lambda x(1-x)$ where $0<x<1$ with $\lambda < 1$ and iterate it [i.e., let each output be the new input to the function so that

$y_1=f(x_1)$,
$y_2=f(y_1)=f(f(x_1))$,
$y_3=f(y_2)=f(f(y_1))=f(f(f(x_1)))$, etc.],

for $\lambda = 0.7$, we would get the graph shown in Figure 6. As we continue the iteration, the iterated output tends to cluster about a fixed point, x^*. In this case, $x^*=f(x^*)\sim0.643$, the intersection of the $y=x$, 45 degree line. In fact, there are two points where the parabola intersects the 45-degree line, $x=0$ and at x^*. Plugging values of either into the equation of the parabola will always give you the same answer. These are both fixed points of f, then. But there is a difference. The slope of f at $x=0$ is greater than 45 degrees while

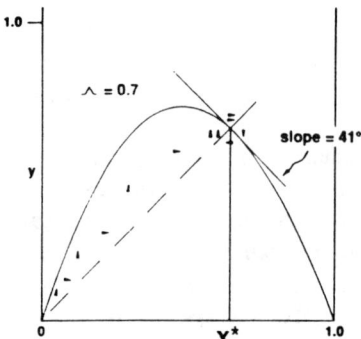

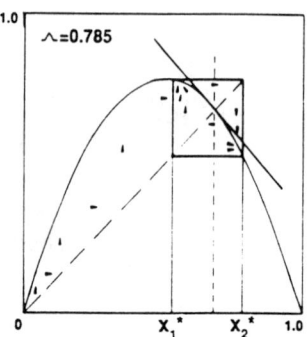

Figure 6. The parabola for lambda=0.7. Iterations define x*, the stable fixed point of period one. [5]

Figure 7. The parabola for λ= 0.785. Iterations give rise to two stable fixed points, x₁* and x₂*. The system has period doubled from that described by Figure 6. [5]

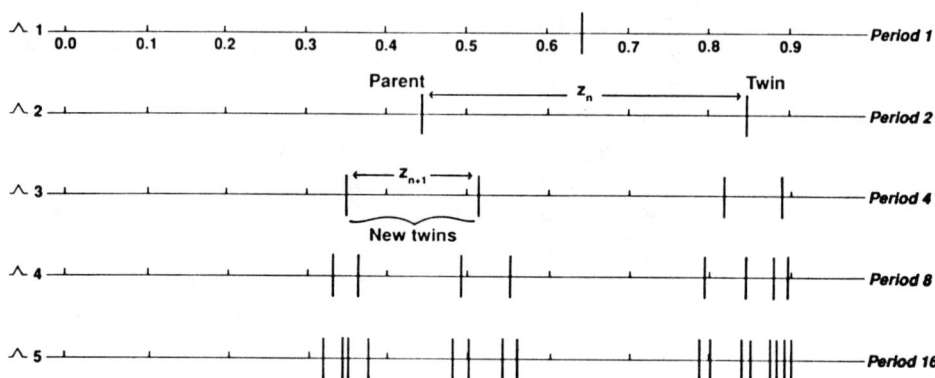

Figure 8. One road to chaos is fraught with period doubling. Each parent conceives two offspring. As n —> ∞, the maximum separation between the offspring obeys Feigenbaum's relationship for α. The horizontal line represents the range of x values for each λ. The vertical lines depict the final values, x*, from iterating the function for that particular value of λ. The separation between each twin grows until it reaches a maximum whereupon each twin gives rise to its own set of twins. (Note: this diagram depicts period doubling. The author lays no claim as to the accuracy of the numbers where doubling occurs.) [5, 6]

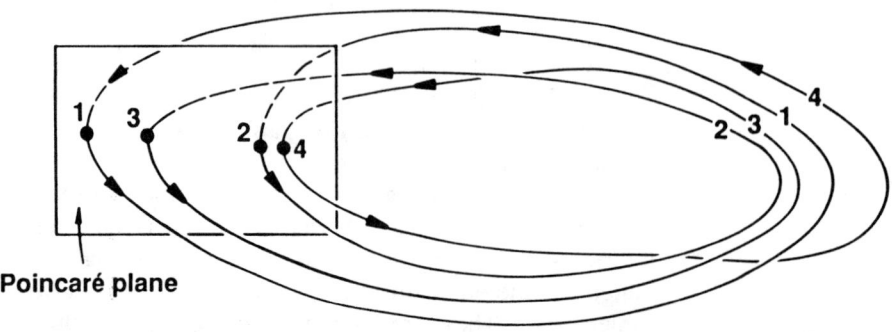

Figure 9. A period 4 orbit cut by the Poincaré plane. The orbit closes on itself after 4 cycles. [6]

the slope of f at x=x* is less than 45 degrees. Any value of x fed into the equation (even those infinitesimally close to x=0 but not x=0) will be drawn or attracted to f(x*) and repelled from x=0. Thus x* is called a stable fixed point and x=0 a repellent or unstable fixed point. That is, x* is a constant solution since each repetition of the iteration gives the same result. And since it repeats after one iteration, it is labeled an attractor of period one.

What happens if we increase the value of λ? It turns out that the slope of the parabola at x* is intimately tied to the choice of λ. For $\lambda = 0.785$, a typical seed value of x may begin to head toward f(x*) but will eventually settle down to oscillate between two other points, $f(x_1{}^*)$ and $f(x_2{}^*)$, as shown in Figure 7. The slope at f(x*) is now greater than 45 degrees and f(x*) has become an unstable fixed point. x* has split into two values, $x_1{}^*$ and $x_2{}^*$, where the solution alternates between the two values. The pair of points, $x_1{}^*$ and $x_2{}^*$, is called an attractor of period two because the solution takes two iterations to repeat. [See 5 for a more detailed discussion.]

What has occurred so far? An attractor of period 1, x*, has split into an attractor, the pair of points $x_1{}^*$ and $x_2{}^*$, of period 2. If we increase the values of λ, we can surmise that this period doubling will occur again. And it does. Figure 8 shows the period doubling through $\lambda = \lambda_5$. Each line is one half of a twin spawned from a single parent (chaotic conception, perhaps?). Moreover, in the limit of infinite period doublings, the distance, z_{n+1}, between the most widely separated of progenies in the (n+1)st generation is α times shorter than the distance, z_n, between those twins' parent and the parent's twin, where $\alpha = 2.502907875...$ is a constant first discovered by Feigenbaum [9]. In turn, the parent and its twin is the most widely separated of progenies in the nth generation* (See Figure 8). In other words,

$$\alpha = \lim_{n \to \infty} \frac{z_n}{z_{n+1}} = 2.502907875...$$

Feigenbaum also discovered another constant which he called δ. As λ is increased, the λ_n's at which period doubling occurs approaches a critical value, $\lambda_c = 0.892486418....$ The distance between successive λ_ns shrinks geometrically. As the number of

---

*In general, the distance between twins in each generation is scaled according to the coefficients in a binomial distribution. For example, at the period 16 level of Figure 8, the expansion is $(a+b)^3 = a^3+3a^2b+3ab^2+b^3 = \alpha^3+3\alpha^4+3\alpha^5+\alpha^6$, where a=$\alpha$, b=$\alpha^2$. Of the eight intervals, three have approximate length of α^4, three have approximate length of α^5, one has approximate length α^3, and one has approximate length α^6.

generations, n, approaches infinity, the ratio of values between consecutive λ_ns approaches the constant, δ. That is,

$$\delta = \lim_{n \to \infty} \frac{\lambda_n - \lambda_{n-1}}{\lambda_{n+1} - \lambda_n} = 4.66920166...$$

The two Feigenbaum constants, α and δ, define precisely a universal path to one form of chaos. They also describe precisely the developing complexity of chaos. And it is above λ_c where chaos's "sensitive dependence on initial conditions" occurs. Also, for $\lambda > \lambda_c$, orbits become aperiodic.

Feigenbaum then tried his iterations on a sine curve. To his surprise, he got the same behavior and the same numbers for the constants. He tried other functions. As long as they had a smooth maximum, the same behavior arose! Here, then, were the first inklings of something that perhaps stretched beyond a mere localized phenomenon.

None of this would have meant anything and would have been filed away in the realm of pure mathematics if it hadn't been for the fact that many researchers, working "in the dark" before the explosion of interest in nonlinear dynamics and chaos, found period doubling in many physical systems in such fields as fluids, lasers, biology, neurophysiology, astronomy, meteorology, oceanography, etc. One of the early researchers, an Italian scientist, Valter Franceschini, University of Modena, who was working on turbulence, found that the system he was studying exhibited attractors and period doubling. He showed his work to Jean-Pierre Eckman, University of Geneva, who suggested he calculate the rate of convergence for the λ values at which period doubling occurs. To their amazement, they found the values matched Feigenbaum's constants!

And yet, Feigenbaum had already felt confident that his relationships would hold up in physical systems. Why? Because he could show that there was a link between phase space trajectories, which described physical systems, and his differentiable nonmonotonic functions, i.e., those functions having a smoothly curved maximum. Poincaré had shown that in a three-dimensional phase space one can invent a plane perpendicular to the phase space trajectories. The trajectories would punch holes through this Poincaré plane (Figure 9). From the pattern of punched holes, one could figure out a relationship and predict where the next hole would be punched.

If we look at Figure 8, the original "parent" defined by λ_1 states simply that an orbit closes on itself. It makes one hole in the

Poincaré plane. For λ_2, the orbit closes on itself only after two revolutions. On the Poincaré plane, it cuts through the plane first at 0.44, then at 0.84, and closes again (or repeats itself) at 0.44— hence the nomenclature, period 2. If we look at λ_3, we see a period 4 system. Let us say the first trajectory cuts the Poincaré plane at 0.35. A possible sequence of trajectories, then, as the orbit loops around and cuts the Poincaré plane, could be 0.35, 0.82, 0.52, 0.88, then 0.35 again. So the orbit closes on itself after four times (hence, period 4), and the cycle repeats itself.

How does that help us discover the function that gave rise to the period doubling shown in Figure 8? Considering λ_3, we begin by plotting the mth value against the (m+1)st, i.e., the location on the Poincaré plane where one trajectory intersects it against the location of where the next trajectory will intersect it for any particular λ. On an x-y graph, the four pairs of points would be: (0.35, 0.82), (0.82, 0.52), (0.52, 0.88), and (0.88, 0.35). These points fall on the parabola. Thus, from the data, we can reconstruct the function! Note that each x-value of the ordered pairs follows the sequence denoted in the paragraph above for where successive trajectories intersect the Poincaré plane.

Suddenly, we have come full circle. From the simple iteration of a function with a differentiable maximum, we have shown a connection to phase space trajectories and vice versa. Suddenly, it seems, mathematics has once again become relevant to real-world physical systems, as Feigenbaum had surmised.

Fractals

A chaotic attractor is a fractal. And what is a fractal? It is an object that exhibits more structural detail as it is increasingly magnified. "If you look at a fractal on any scale, you will start to see new structure that is statistically similar (self-similar) to the structure of the whole," Robert Brammer of The Analytic Sciences Corp. explained.

"A fractal is something in which what looks like a line on one scale of magnification, when magnified more, turns out to be a series of nearly parallel lines with gaps in between. When you magnify each line, it becomes a series of lines," Abraham said (Figure 10). "And any surface that you may see is not, in fact, a surface, because when you magnify it, it turns out to be layers of surfaces. And each one of those layered surfaces, when magnified more, is layers of surfaces. The fractal itself is an object that can never be pinned down to a smallest structure in the sense that a mathematical point or line can be pinned down, because a

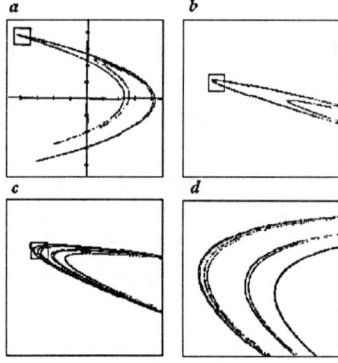

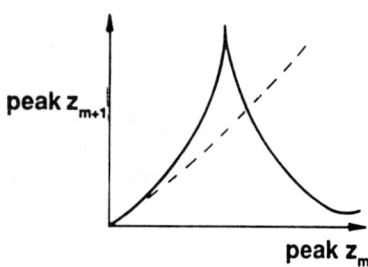

Figure 10. The fractal nature of chaos is evident as we continually zoom in on parts of a path of a strange attractor. Magnification reveals more and more details that are self similar to the previous, more macroscopic pictures. [2] (Courtesy of James P. Crutchfield)

Figure 11. The cusped function that Lorenz found is a sign of intermittency chaos, i.e., there is no period doubling to chaos. z is the function defined in Table 1. A plot of z versus time gives rise to a spectrum of peaks. Each peak plotted against its adjacent peak gives rise to the cusp above. [6]

mathematical point has zero dimensions and a mathematical line has zero thickness. They do not change when magnified."

And what about period doubling? Is that the fractal nature of chaos? When you stand far enough away from a period doubled system, you would only see two blurry lines. But as you get closer to each line, you would see that, in fact, each line is made up of two blurry lines with a gap between. Suppose you are at the level of period 16 (λ_5, Figure 8). One can see structure on magnification by a factor of 2 for four magnifications. But after the last of those magnifications, the lines become sharply delineated. Thus, in this period doubled region, a fractal does not exist. The fractal appears only after one goes through an infinite set of period doublings, i.e., for λ greater than λ_c.

The road to chaos

When a system surpasses the value for λ_c, period doubling has crossed the path into chaos. Now, there is no recognizable periodicity to the system. Trajectories have that sensitive dependence on initial conditions and diverge on a short time scale to reconverge and diverge again. And then there is the mixing and folding of trajectories, but the trajectories do not intersect.

But there are also other roads to chaos. These include Lorenz's original path, which did not include period doubling, but rather what is called intermittency. By plotting the value of the (m+1)st peak of one of the variables against the value of the mth peak, he came up with a cusped function (Figure 11). "This cusped map is the signature of intermittency chaos and indicates that there would not be period doubling, and none was observed," Abraham said. Remember, a period doubling system gives rise to a function or map (such as a parabola or sine function) with a differentiable (smoothly curved) maximum.

"And other forms of chaos come from the Ruelle-Takens route known as quasiperiodicity, which has several variations," Abraham said.

So the road to chaos is a branched one. But the end result is still chaos.

The future

James Gleick, in his book *Chaos*, subtitled it appropriately enough as *Making a New Science*. Not since the early days of this century, with the heady discoveries in quantum mechanics and relativity, has there been as much interest and excitement in physics and mathematics. The study of chaos has loosened and redefined the straitlaced, rigorous and conventional, proof-as-principle definition of mathematics to include, within its domain, computers and computer graphics to handle large numerical data sets and repetitive operations. When Feigenbaum showed his results at one meeting, the noted mathematician Mark Kac brought up the question of whether Feigenbaum's talk would be based on mathematical numbers or mathematical proof. More the former than the latter, Feigenbaum answered. "Is it what any reasonable man would call a proof?" Kac then asked. Each listener would have to judge for himself, Feigenbaum replied. After hearing Feigenbaum's results, Kac said: "Yes, that's indeed a reasonable man's proof. The details can be left to the r-r-rigorous mathematicians."[4, p. 184]

Moreover, the study of chaos has cut across traditional boundaries. It is not confined to just physics and mathematics. It is also being used in meteorology, biology, neurophysiology, economics, astronomy, fluids, and other fields. In fact, anywhere where processes that had been previously thought to be random will now bear new investigation.

One could always say that there is chaos in the beauty of nature. In a Miltonian sense, who cannot begrudge the terrible, chaotic beauty of a lightning storm on a midsummer's night? But, now, we can also say that there is beauty in chaos, because with our new definition, we have a new tool for understanding more of nature's roiling processes. As Feigenbaum said, "I truly do want to know how to describe clouds... Somehow the wondrous promise of the earth is that there are things beautiful in it, things wondrous and alluring, and by virtue of your trade you want to understand them." [4, p. 187]

References

1. Abraham, Neal B., "A New Focus on Laser Instabilities and Chaos," *Laser Focus*, May 1983, p. 73-81.

2. Crutchfield, James, J. Doyne Farmer, Norman Packard, and Robert Shaw, "Chaos," *Scientific American*, December 1986, p.46+.

3. Fisher, Arthur, "Chaos: The Ultimate Asymmetry," *Mosaic* (National Science Foundation), Jan./Feb. 1985.

4. Gleick, James, *Chaos, Making a New Science*, Viking, New York, NY, 1987.

5. Hofstadter, Douglas, "Metamagical Themas—Strange attractors: Mathematical patterns delicately poised between order and chaos," *Scientific American*, November 1981, p. 22+.

The author would like to thank the following people for their helpful discussions:

6. Neal Abraham, Physics Department, Bryn Mawr College, Bryn Mawr, Pennsylvania.

7. Robert Brammer, The Analytic Sciences Corporation, Andover, Massachusetts.

8. Robert Gioggia, Physics Department, Widener University, Chester, Pennsylvania.

9. John Guckenheimer, Mathematics Department, Cornell University, Ithaca, New York.

10. Shirly Dawson, Department of Radiology, St. Joseph Hospital, Bellingham, WA.

11. Jerrold Marsden, Department of Mathematics, University of California, Berkeley, CA.

Further suggested reading:

12. Cvitanovic, Predrag, ed., *Universality in Chaos*, Adam Hilger Ltd., Bristol, 1984. This book is a compendium of some of the classic papers.

13. Frisch, Uriel and Steven Orszag, "Turbulence: Challenges for Theory and Experiments," *Physics Today*, January 1990, p. 24+.

14. Goldberger, Ary, David Rigney, and Bruce West, "Chaos and Fractals in Human Physiology," *Scientific American*, February 1990, p. 42+.

Neal Abraham, Robert F. Brammer

Optical Chaos

Neal Abraham

Robert F. Brammer

Neal Abraham is Rachel C. Hale Professor of Mathematics and the Sciences, and professor of Physics at Bryn Mawr College (Bryn Mawr, Pennsylvania). His fields of specialization include laser physics, quantum optics, nonlinear dynamics and chaos, and stochastic fluctuations in optical sources. He is a Fellow of the Optical Society of America and has taught at numerous universities around the world.

Robert F. Brammer received the BS from the University of Michigan and the MS and PhD from the University of Maryland, all in mathematics. He has worked for the NASA/Goddard Space Flight Center on the Apollo and Skylab projects. In 1974 he joined The Analytic Sciences Corp. (Reading, Massachusetts), where he is vice president and technical director of the Strategic Sciences Group.

Neal, why is chaos and nonlinear dynamics important in nonlinear optics?

Abraham: An understanding of chaos in nonlinear dynamics gives us a whole new outlook on the behavior of optical systems. While much of our classical understanding of optics deals with noise from spontaneous emission, mechanical jitter, or incoherent emissions from extended sources, an understanding of chaos tells us that not all irregular signals arise from randomness; some irregular signals may be chaotic and there are ways to measure characteristics of these signals to see if they are chaotic. If the chaos is undesirable, then very different measures must be taken to remove it than would be taken to remove random noise. Changes in deterministic chaos would require changes in the dissipation rates or in the nonlinear interactions, while removal of random noise would require decoupling of the system from the sources of the noise. Chaos and complex nonlinear dynamics have been found in mode-locked lasers, single-mode lasers, pulsations from phase-conjugate mirrors, the interaction of counterpropagating beams, bistable optical devices, and in many other situations.

Neal, what can you tell us about chaos as it relates to lasers?

Abraham: Well, there are many different lasers that are either spontaneously generators of chaotic signals or do so when they are

The discussion was conducted by Frederick Su. It is reprinted from *OE Reports*, April 1990.

used in certain applications, such as with external feedback mirrors designed to narrow their linewidth or with modulation designed to generate signals for communications. Lasers in many of those situations or applications (lasers which are modulated or used with external feedback mirrors) may generate chaotic signals rather than the simple, pure-modulated or narrow linewidth signals that the designs were intended to create. So those are instances where chaos is a form of signal degradation. It is called, for example, "coherence collapse" in semiconductor laser terminology. What happens is that the temporal coherence of the laser disappears; in fact, the feedback methods that were being used have suddenly gone haywire. It's like in any feedback loop when the feedback is too strong and the system has gone into a nonlinear regime in which it is generating its own spontaneous noise, as it were, which is not noise in the most generic sense, but is in fact chaos.

Bob, you are studying chaos and fractals in terms of image computing. Why don't you give us some background on that?

Brammer: I think it is important at the outset to distinguish between the dynamical aspects of chaos modeling and fractals, which are geometrical objects which can often be generated by nonlinear dynamical systems. So when we are talking about chaos, per se, we tend more to think about dynamics, and when we think about fractals we tend to think more about geometry; the two concepts are very closely interlinked. The thing that makes them particularly important in the field of image computing is the fact that these nonlinear dynamical models of images or of scenes from which these images are made are fairly general. They apply to many situations, particularly those of natural scenes, such as terrain, clouds, and other natural objects. Benoit Mandelbrot and others have shown that fractal models are very good descriptions for many natural scenes. What we have found in our research is that chaos models of fractal objects and fractal modeling techniques apply in all areas of image computing. That is, these models can be used in certain types of image analysis, image communications, and image synthesis problems. So it affects all areas of this field.

Does it help, Bob, to have an understanding of the dynamical chaos that gives rise to these fractal geometries for you?

Brammer: In certain cases it does. For example, the work by Barnsley and colleagues at Georgia Tech uses a dynamical model to synthesize certain classes of images. One can efficiently make, in certain areas, very realistic images with these types of chaotic dynamical models.

Figure 1. Fractal concepts are used in cloud modeling as seen in this picture enhanced from data captured by the Geostationary Operational Environmental Satellite.

What are they modeling exactly?

Brammer: They can model a wide variety of natural objects, photographs, etc. Where it tends to work best are in the natural scenes.

What about weather modeling?

Brammer: That is a particularly good example. Chaotic models and fractals are used in various aspects of weather analysis forecasting. In fact, the roots of the whole field of chaos really came out of meteorological forecasting work of Professor Lorenz at MIT back in the late 50s and early 60s. Under certain circumstances the atmospheric phenomena are turbulent and chaotic. So the models that are used to forecast certain types of weather phenomena are exactly those nonlinear dynamical models that lead to certain types of chaotic behavior. But it does not stop there. Clouds are a good example for the modeling of atmospheric phenomena (Figure 1). There have been some very simple dynamical models of clouds developed. The work of Grassburger and Procaccia is a good example of this. Their nonlinear dynamical models of the internal dynamics of cloud systems actually led to predictions that were verified with, for example, satellite cloud imagery. There has been some empirical work by Lovejoy and Mandelbrot about the ratio of cloud areas to cloud perimeters in satellite imagery, and with a fairly simple fractal model it was determined that the fractal dimension of a cloud perimeter is about 1.35. That is a noninteger dimension. Not like a line or a surface, but a fractional dimension. These dynamical models actually led to a prediction very close to that which had been observed empirically in the measured satellite imagery. So there are some very tight connections between the nonlinear dynamical models and the fractal geometrical descriptions that have shown up in several parts of weather analysis and forecasting.

Neal, doesn't cloud formation, since it is a model of turbulence, raise in your mind the problem of sensitive dependence on initial conditions and why weather forecasting is good only for a short term?

Abraham: I think there are several different things here at once. Let me perhaps first say that there is a link between chaos and fractals of a different sort, which is one separate from the patterns that a chaotic process might generate in space, a topic we will get back to. In its own phase space, a chaotic system is occupying a fractal subset and so if one is able to measure something which is a single variable changing in time, and if one can construct from that a phase space portrait, then that phase space object is itself a fractal. So that part of what chaos studies involve is the reconstruction and then measuring of the fractal, which is taken to be a signature of the existence of a chaotic process that is generating the signal. When one gets into processes which generate fractals in real space, then one clearly is talking about turbulence, spatial-temporal phenomena rather than just temporal phenomena, and those are not quite properly called chaotic per se. One must solve the partial differential equations of variations of space and time. And so the generic name that is applied to that is more generally turbulence rather than chaos. There is evidence that turbulence, for example, has small little cells, each of which behave chaotically, but that the cells are less and less correlated with each other as one moves farther and farther apart within the overall turbulent system. To have that whole process in space and time generate a fractal pattern in real space is not something you can connect directly to a positive Lyapunov exponent, because it is hard to be sure what you are defining when you want to talk about the Lyapunov exponent for a system that varies both in space and in time.

The evidence is that if you have sufficiently bounded turbulent systems in the various laboratory fluid experiments, you can make fluid dynamics problems begin to behave chaotically. And if you make them too big, then different parts of the experiment seem to be independent of each other. One can measure at any one point, calculate properties of the chaos, and then try to find correlations between different points in the pattern. One loosely uses expressions such as, "Positive Lyapunov exponents of chaos explain why the weather is not predictable." Well, in its most precise form, such a statement assumes that you can write down one single set of ordinary differential equations to describe the weather, and I don't know that anyone is so bold as to suggest that. We are not quite sure what to call these phenomena when they are on such a large scale. That is really at the forefront of studies both in optics, such as for large-aperture laser beams where you have a large Fresnel number, or for large realistic fluid flow systems. One is there at the forefront trying to understand how you would characterize the dynamical properties of an extended system, optical or fluid, that is varying in both space and time.

Neal, I want to go back to the laser systems. In chaotic laser systems, is it a period doubling route to chaos or is it the Ruelle-Takens route to chaos, which we barely touched on last time?

Abraham: Almost every known route to chaos has been observed in lasers. In fact, many types have been observed nowhere else, that is to say, ones have been observed that are not yet universal enough to have names. There is no careful rhyme or reason why particular systems become chaotic in particular ways when you come at them from their physical side rather than from their mathematical modeling side. Certain cases are simple enough that one finds Ruelle-Takens routes to chaos. One has a two-frequency route to chaos, particularly in systems that have their own internal spontaneous modulation frequency and then which are modulated at a different frequency. Other systems, both modulated and spontaneously pulsing, will show at least partial period doubling sequences. Others undergo various other kinds of bifurcations. There are those that undergo an intermittency form of chaos such as what the Lorenz model displays. There are lasers which emulate that quite closely. There are others which show something known as homoclinic chaos. So there is a very rich variety. The easy ones, from an experimentalist's point of view, are those which do show period doubling or a Ruelle-Takens type pattern because those are easy for relatively inexperienced experimentalists to say, "Oh yes, I have read something about that. Now I know where to go to try to follow up on this." On the other hand, there are lots of others that just go nuts. As when the semiconductor laser people try to make narrow linewidth semiconductor lasers by adding extra feedback mirrors. They suddenly discover that the laser, instead of having a narrower linewidth, has gone up in linewidth by a factor of 10 or 100. There is a sudden spontaneous jump to a broad linewidth instead of the narrow linewidth that was sought. No one has identified that route; it is just a spontaneous switch-over from stable behavior to chaotic behavior. One really cannot generalize to say laser chaos is of a particular type. Different lasers have sufficiently different nonlinearity or different systems of mirrors, feedback, and modulation to cause different couplings.

Can you elaborate quickly on the Ruelle-Takens route? We touched briefly on it before in the "Chaos" article.

Abraham: This occurs in laser physics for some lasers with intracavity saturable absorbers. Two examples are gas lasers with appropriate molecular gas absorbers and semiconductor lasers that have been damaged or that have sections which are insufficiently excited so that they act as absorbers. Lasers of that type can spontaneously generate and sustain a pulsing pattern; they are emitting a single optical frequency but the intensity is also pulsing. If one modulates the gain or loss of such lasers with another frequency, then the overall output is modulated with two relatively independent frequencies; one is the internal frequency of modulation and the other is the external modulation frequency. Since the internal modulation frequency is generated within the dynamics, the fact that you are forcing from the outside (much as when one forces a pendulum or forces a metronome or pushes a child on a swing) means that to some degree you are forcing the system to oscillate at the forcing frequency. And if you drive the system hard enough, one of the things that can happen, which is part of the Ruelle-Takens route, is that the internal frequency locks to the external frequency. It might lock at exactly the same frequency or it might lock at a harmonic or subharmonic. They lock into some rational ratio which is a way of forming a kind of synchrony—and after some finite length of time the pattern begins to repeat again. Now, in the Ruelle-Takens route several different things can happen. After you have a locked state of the two frequencies, there might be a period doubling to chaos. Or, if the frequencies do not lock and the system persists in having two independent frequencies which have an irrational ratio, there might be a sudden breakdown into broadband (many-frequency) chaos. The rule of thumb often is if a third irrationally related frequency were to appear in the pattern, then the system is likely to be very unstable and make the transition immediately to broadband chaos. That led to some interest for a while in trying to see whether anything existed which had three independent frequencies. And it is not hard to rather arbitrarily construct systems which have that. It is not an absolute rule that three incommensurate frequencies (where the ratio of one frequency to another is an irrational number) will generate chaos, but it has been shown by several people who have studied this route in detail that the probability of a system being able to sustain three nonlinearly coupled incommensurate frequencies of modulation is really quite low. In such a situation the system is most likely to make a transition into chaos. The basic route, then, is to have first one frequency, add another independent frequency, then most typically to lock those frequencies together and have that locked state go into some form of chaos, either by period doubling or by directly breaking into chaos.

Bob, in terms of your image computing and image analysis, what information and what advantages does chaos and fractal modeling afford us.

Brammer: In some areas, I think a major advantage is that a compact efficient description of the phenomena can be formed. For example, the iterated function systems that have been used in certain areas of image synthesis or the fractal models that have been used for object detection image analysis give you a simple way of describing your phenomena and a compact way of communicating it. A good example of that is the use of fractal dimension as a segmentation technique in image analysis. As you scan across the image-calculating fractal parameters, such as the fractal dimension in various sub-areas of your image, you can segment the image according to differences in fractal dimension (Figure 2). This can be done automatically, entirely by computer without human intervention, and it can be very efficient. So, you can use that as a

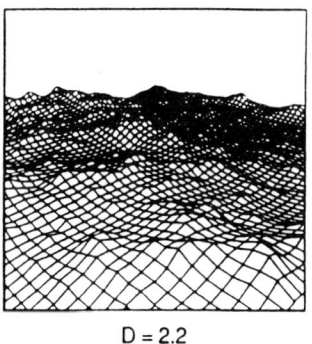
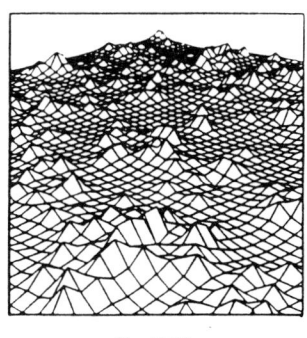

D = 2.2

D = 2.85

Figure 2: These simulated fractional Brownian motion-surfaces are fractals generated from the random walk. Statistically, the particular path of Brownian motion has random jumps on all scales, and is thus self-similar at all magnifications. Note that the higher fractal dimension implies a rougher texture for the surface. Thus, fractal dimensions provide a method to segment or separate out parts of an image.

In the case of random noise, one would have a kind of diffusive random walk of the two solutions away from each other, but they would not explode away from each other exponentially. The exponential sensitive dependence on initial conditions is a unique feature of chaos.

way of discriminating man-made objects against natural backgrounds. Man-made objects have entirely different dimensional characteristics in images than do natural objects such as trees, rocks, or other natural things. To answer your question, a major advantage of the use of chaos models, fractals, and related aspects in image computing is the fact that they provide a compact and efficient model, which can be used to do calculations in a highly automated way.

Also, you can build up with fractals the most complicated type of scene imaginable, right?

Brammer: That's right, generally by iterating you can generate scenes of significant detail again, but in a very compact code, so you don't need a lot of complexity in your description in order to generate a lot of complexity in your scene. The use of the compact and efficient model can lead to a very complex object if that is important for your application.

Great. Neal, back to you. You talked about reaching chaos in these laser systems, but from what we talked about before, how do you identify, how do you know that a laser has turned chaotic?

Abraham: One of the simplest and least precise of the signatures of the onset of chaos is the appearance of a very broad band of frequencies in the spectrum of the system (Figure 3), much as spontaneous emission noise would appear as a broadband spectrum. The unfortunate part is that the appearance of a broadband spectrum is then a less-than-optimum indicator that chaos is the cause. If one can record the time-dependent behavior, then one has the signal which can be used to reconstruct an attractor or reconstruct the behavior in the phase space of the system. That behavior has several features which can be identified with chaos. One of those is that the attractor has a fractal dimension (Figure 4), so one would use programs much like those for image analysis of clouds or other fractals to analyze, in this case, the phase space portraits of the system.

There are a few cases, at the moment not enough to be worrisome, where one has been able to show that fractal phase space portraits are generated by systems that have multiple periodicities and not, in fact, chaos. But that said, examples are still so rare that one can rather generally say that if one finds a fractal dimension for the attractor of a signal, then one has rather good confidence both about the existence of a chaotic origin for that signal and about the number of key variables needed to describe the dynamics. Another feature that could be calculated would be something known as the Lyapunov exponents, which is how rapidly the solutions of the system diverge when you begin with neighboring initial conditions. You need to have a signal that is long enough so that you can find two starting points that are the same and to see if they diverge exponentially. If they do, that is another signature of chaos as opposed to random noise. In the case of random noise, one would have a kind of diffusive random walk of the two solutions away from each other, but they would not explode away from each other exponentially. The exponential

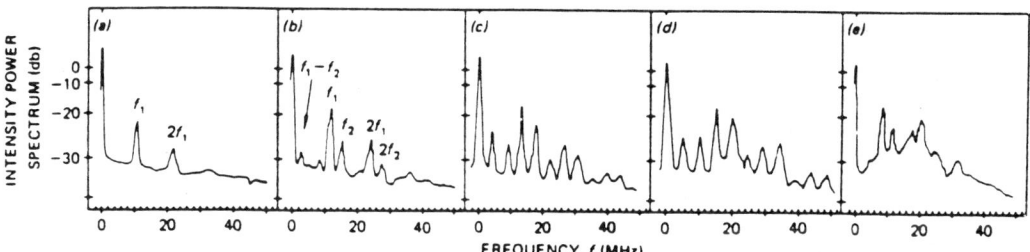

Figure 3: A sequence of intensity power spectra of a spontaneously pulsing xenon laser illustrates the Ruelle-Takens route to chaos. (a) Periodic, (b) Quasiperiodic (two incommensurate frequencies), (c) Locked at a ratio 4:3, (d), (e) Emerging chaos indicated by broadband (blurring) of frequencies.

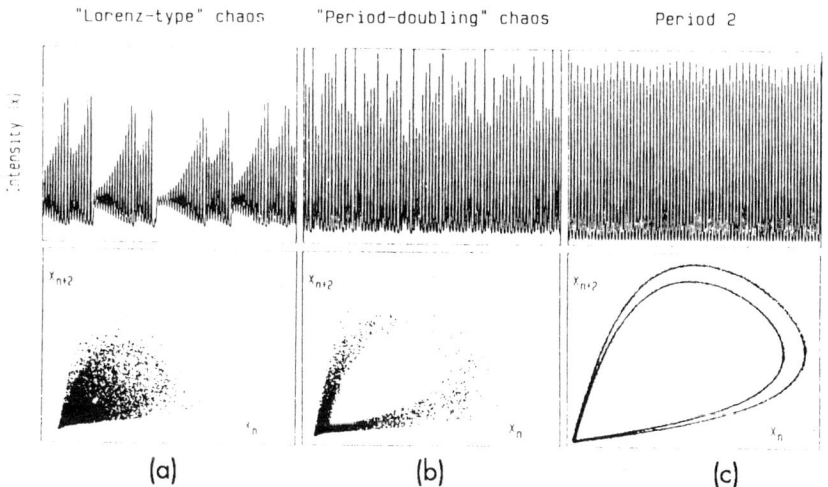

Figure 4: The top row shows intensity pulsations from an ammonia laser. Each column represents different operating conditions. The second row shows projections of the reconstructed phase space attractor. (a) and (b) are chaotic, (c) is periodic. The scattered points in the second row for (a) and (b), the blurry regions, lie on a fractal of about dimension 2.1. (c) The periodic signal is revealed by two loops, which indicates a one-dimensional attractor. (Anything periodic is one-dimensional).

The use of these models—chaos models, fractals, and related techniques—I think, offers a class of models that can be effective at least in certain applications for achieving compact descriptions of images. And to a fairly high level of accuracy, you can then operate on the description rather than every pixel in the image.

sensitive dependence on initial conditions is a unique feature of chaos. So both the Lyapunov exponents and dimensions are two of the easy things to measure. A third is something related to information rate—how much information you need to describe the signal. If it is random at some rate given by its bandwidth, you need absolutely brand new information every inverse of the correlation time in order to tell what is going to happen next. If it is deterministic, that information is needed at a somewhat lesser rate. If it is periodic, then once you have tracked one period of the system you are able to forecast the whole future without needing any additional information. So this "information theory" type of approach leads to the definition of entropies. The entropy rates characteristic of periodic signals are zero; the characteristic of chaos is a finite entropy rate while the characteristic of truly random white noise is an infinite entropy rate. So one has a range of possible values of entropy or information rate that can be calculated as a third feature of the chaotic signals. It is easy with some optical signals because the pulsation frequencies are in the kilohertz or megahertz range. But for many of the technologically important lasers such as semiconductors, the pulsation frequencies are gigahertz, and it is hard to record such signals at the necessary 10 or 20 gigahertz sampling rate with enough resolution and length to be able to reconstruct things. We are right at the edge of electronic technology in being able to record signals from some of the very interesting cases that we would like to analyze for the presence of chaos. If we do, we may find that it is not chaos that is causing the noise, but some other breakdown that leads to an internal noise source related to spontaneous emission or some other process.

Bob, what do you see in the future in terms of image analysis, image computing, since it seems that we are just starting to scratch the surface on fractal analysis and chaos?

Brammer: I think a very important area for further research is in the area of image modeling. We don't really have enough general descriptions of the types of phenomena that appear in images that are commonly used for scientific, engineering, or even commercial applications. As such, we tend to treat most of these images empirically. The use of these models—chaos models, fractals, and related techniques—I think, offers a class of models that can be effective at least in certain applications for achieving compact descriptions of images. And to a fairly high level of accuracy, you can then operate on the description rather than every pixel in the image. That means you have the potential for automating certain types of analysis, of speeding up or making the image communication more effective. Then, if you need to, you may be able to accurately regenerate the image at some remote location without having to transmit the entire image. So I think, at least for certain classes of images, there is a trend toward the use of these models in unifying these areas of image computing. I believe that it is an important trend. I expect new developments in that area will continue throughout the 90s.

Neal, I want to finish up by asking you how important you think chaos theory is in the general study of nonlinear optics.

Abraham: I think there is increasing evidence that as we press to higher and higher nonlinearities to generate more interesting special effects, such as phase conjugation or photorefractive effects, we are finding that the more complicated forms of nonlinear dynamical behavior will appear. We either have to be prepared to find interesting uses for them, or we have to be prepared to recognize that chaos has appeared and take appropriate steps in terms of modifying the dynamics in order to suppress it if one wants to have stable nonlinear behavior rather than the chaotic nonlinear behavior. The theory needs to keep up in order to tell us how to recognize chaotic behavior, to explain how to modify chaotic behavior to either periodic or stable phenomena, or to tell us that chaotic signals have certain uses that will be particularly valuable in doing things that need to have fluctuating signals rather than constant signals. I think it can go both ways, that we will both find uses for chaos as we better understand its properties and better be able to get rid of chaos when we understand its origin.

Anything else to add?

Brammer: I would like to ask Neal a question. Is there any part of your work that deals with the propagation of nonlinear pulses, solitons and so forth, through long-distance fiber optics? And is there any relationship in these nonlinearities to chaotic phenomena?

Abraham: There are at least two links. There have been those who have been doing nonlinear mathematics, particularly studying nonlinear partial differential equations, who have been working in both fields; they have on occasion called us together to talk at least near each other, if not at each other. There was a meeting at San Jose State University in January 1988, I guess, that was on solitons and chaos. (Solitons are localized pulses of energy that retain their shape as they propagate in space or nonlinear materials, and they retain their integrity as they collide with other pulses). More generally in the nonlinear optics field, solitons are generated for a variety of signal communications problems. Certain lasers are specially designed to generate solitons. And, fancy kinds of transmission lines are designed to maintain what are nearly solitons by providing a little bit of special amplification to offset the little bit of loss over long-distance fibers. Also, some transverse patterns on laser beams in nonlinear materials also seem to have the character of solitons.

Brammer: I think this is an important subject that I wish we all knew a little more about. When we think about image communications, the use of wideband fiber optics is obviously very important, and the use of chaotic models in transmitting data at high speed through fiber optics, I think, is a very interesting research area.

Abraham: I know a few people who are working on using chaotic signals as carriers or as modulators of various types, but that is relatively more limited than the very widespread work of propagation stability of solitons themselves. I haven't yet seen what would be a significant breakthrough that would be a full-fledged application of chaotic signals in nonlinear optics. We're still discovering new forms of chaos and we know more are there. But their applications are still limited as far as I can tell.

Kip Thorne

Laser Interferometer Gravitational Wave Observatory

Kip S. Thorne received his BS from Caltech in 1962 and his PhD from Princeton in 1965. He returned to Caltech as Associate Professor in 1966, was promoted to Professor of Theoretical Physics in 1970 and became the William R. Kenan, Jr., Professor in 1981. Thorne's research has focused on gravitational physics and astrophysics, with particular emphasis on black holes and gravitational waves. With John A. Wheeler and Charles W. Misner in 1973 he coauthored the textbook Gravitation, *from which most of the younger generation of physicists and astronomers have learned general relativity theory. He has developed much of the mathematical formalism by which astrophysicists analyze the generation of gravitational waves by highly nonlinear cosmic sources; and he has worked closely with experimenters on new technical ideas and plans for gravitational wave detection, including the Caltech/MIT 4-km laser interferometer gravitational wave detection system. Thorne and his students and colleagues have developed much of the modern theory of interaction of black holes with complex astrophysical environments. These researches recently culminated in the publication of their book* Black Holes: The Membrane Paradigm. *Thorne was elected to the American Academy of Arts and Sciences in 1972 and the National Academy of Sciences in 1973. He was awarded an honorary doctorate by Moscow State University (USSR) in 1981. He has been a Woodrow Wilson Fellow, a Danforth Foundation Fellow, a Fulbright Fellow, and a Guggenheim Fellow; and he has served on the International Committee on General Relativity and Gravitation, the Committee on US-USSR Cooperation in Physics, and the National Academy of Sciences' Space Science Board.*

The interview was conducted by Frederick Su.

One often hears about "windows on the universe," and of course the electromagnetic (EM) window, from gamma rays to radio waves, is the one that is normally used. What new knowledge do you think we can gain from studying gravitational waves?

The sources of gravitational waves as predicted by general relativity should be extremely different from the sources of electromagnetic waves. Whereas cosmic electromagnetic waves

Reprinted from *Optical Engineering Reports*, July 1987.

are almost always an incoherent superposition of emissions from individual atoms or molecules or high-energy particles, gravitational waves should be generated by the coherent bulk motions of large amounts of matter, or coherent vibrations of curved empty spacetime, as, for example, when two black holes collide. I can't imagine two emission processes that are more different than these. Correspondingly, the kinds of things one will see with gravitational waves will be very different from what one sees with electromagnetic waves.

What are some other examples besides two black holes colliding?

Other examples of gravity wave sources are the gravitational implosion of a star to form a black hole, or a supernova, in which the core of a star implodes to form a neutron star and the gravitational energy released in that implosion drives an explosion of the star's outer layers. In either of these sources, black hole birth or supernova, the region from which the gravitational waves would come would be the region of very strong gravity, deep in the interior of the imploding star, a region from which it is hopeless to get any electromagnetic waves directly. Electromagnetic waves all come off the star's surface, where the action is not. The gravitational waves come from the core, where the bulk of the action occurs. Even neutrinos can't come out directly from the core of a newly forming neutron star. Neutrinos will scatter thousands of times on the way out, losing most of the information they began with. Gravitational waves are the only things that can get out unimpeded, with no loss of information.

Another source is the big bang at the beginning of the universe. For the first several hundred thousand years of the universe's life, its gas was so hot and ionized that it was opaque to electromagnetic waves. Consequently, when astronomers look at the early universe using electromagnetic waves—the cosmic microwave background— they see it as it was at an age of a few hundred thousand years. Similarly, if you were able to see the early universe by looking at the stochastic background of neutrinos that it presumably produced, you would see the universe as it was at an age of about one second. Any earlier than that the temperature of the primordial gas was so high that it was opaque to neutrinos. By contrast, when one calculates using general relativity and quantum theory the cross section for interaction of gravitational waves with hot matter, one finds that any gravitational waves from the big bang would last have interacted with the primordial gas when the universe was 10^{-43} seconds old—which is the epoch at which the initial conditions of the universe were set, the so-called "Planck epoch." At that epoch,

If you were able to see the early universe by looking at the stochastic background of neutrinos that it presumably produced, you would see the universe as it was at an age of about one second.

"space" and "time" as we know them did not even exist. In their place were some sort of quantized, probabalistic versions of space and time. It's one of the holy grails of present-day theoretical physics to understand the quantization of space and time or, put in other language, to understand the quantization of gravity and its unification with the other fundamental forces. It was in the epoch of that unification, the epoch of quantum gravity, of quantum spacetime, that the primordial gravitational wave background would have been emitted. And so if the gravity wave background could be detected, we might get from it our only possible direct glimpse of the initial conditions of the universe.

Weren't there detectors built to look at gravitational waves in the late 60s and early 70s, of which the most famous was that built by Joseph Weber at the University of Maryland?

In fact, Joseph Weber pioneered the field singlehandedly. He had the vision in about 1959 to recognize that it would be technologically possible to build detectors that could get to interesting levels of sensitivity, where one might actually see something. And singlehandedly, with the rest of the world not paying much attention, he developed the first gravitational wave detectors in the 1960s. These were resonant bar detectors, large bars of aluminum whose fundamental mode of vibration would be driven by a passing gravitational wave, with a readout system

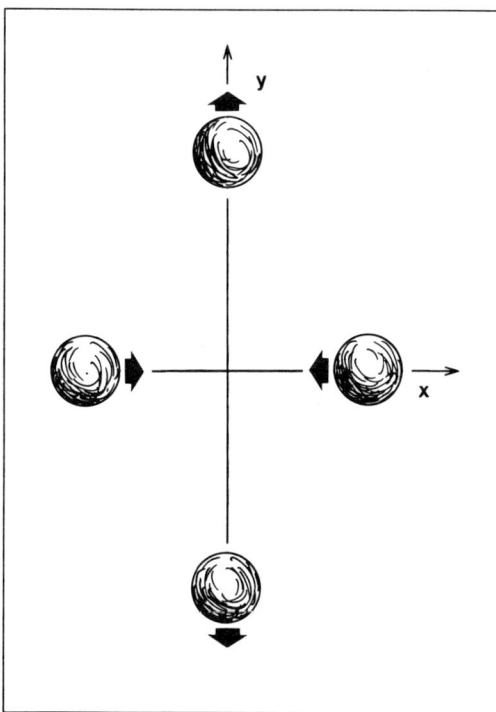

Figure 1. Effect of a gravitational wave on four rectilinearly aligned masses. Wave is into page. Masses along the y-axis move apart while masses along the x-axis move together.

It's one of the holy grails of present-day theoretical physics to understand the quantization of space and time, or, put in other language, to understand the quantization of gravity and its unification with the other fundamental forces.

that looked for sudden unexpected changes in the amplitude or phase of vibration of that fundamental mode. Weber developed a very sensitive readout system based on piezoelectric crystals, and thereby was able to work his way down toward interesting levels of sensitivity. In the late 60s, Weber saw excitation in these detectors, excitations that he interpreted as possible evidence for gravitational waves. These first generation detectors reached a culmination in the mid-70s, at which point other experimental groups—who had moved into the field in the early 70s— were by and large not confirming Weber's excitations. With the culmination of these first experiments a number of groups, including Weber's, started developing a second generation of bar detectors for gravitational waves—bars made of new materials, cooled to liquid helium temperatures, with new kinds of readout systems. In parallel, other groups developed a radically new type of gravitational wave detector, a "laser interferometer gravity wave detector," which I hope will lead in a few years to a "laser interferometer gravity wave observatory." Thus, there has been a very vigorous effort continually since the 1960s on improving the technology, pushing the sensitivity of these instruments higher and higher—an effort carried out by more than 100 very dedicated and very talented experimenters.

Just what are gravitational waves? How do they arise, how do they propagate, and how do they differ from EM waves? What is their

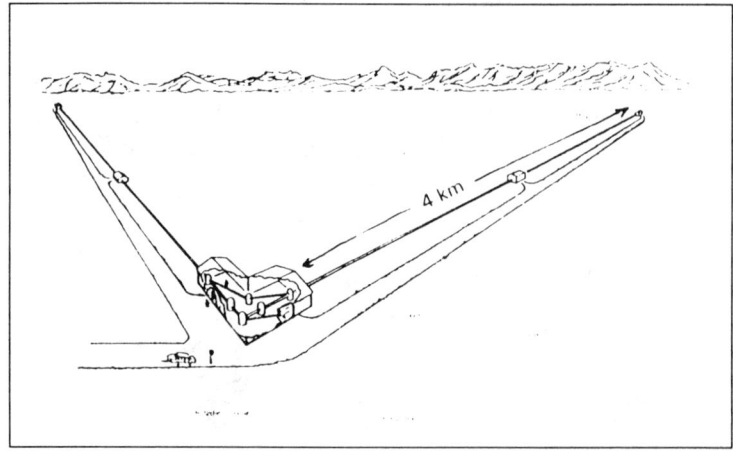

Figure 2. Artist's concept of the the western site of the American LIGO (Laser Interferometer Gravity Wave Observatory), at Edwards Air Force Base, California. The eastern site, in Columbia, Maine, would be similar—though with desert replaced by blueberry bushes and mountains replaced by forests and bogs.

polarization, and how are they detected?

Gravitational waves are one aspect of the gravitational field as predicted by general relativity. If we go back to the electromagnetic case, you'll recall that the electromagnetic field has several different aspects. It includes the electric field, by which static charges attract each other; the magnetic field, by which moving charges are deflected; and wave fields, by which oscillating electric and magnetic forces propagate at the speed of light. Similarly, general relativity predicts that the gravitational field should have four aspects, rather than the three of electromagnetism. It should have an analog of an electric field, which is the ordinary gravitational acceleration by which the sun attracts the planets and the earth holds us to its surface. There should be a gravitational analog of the magnetic field, the "gravitomagnetic field," which would deflect moving masses. There is a planned experiment by NASA—Gravity Probe B—which will search for the earth's gravitometric field through the precessional torque it exerts on earth-orbiting gyroscopes. General relativity says that the third aspect of the gravitational field is the curvature of space, which shows up very clearly in the measurements of the gravitational deflection of light by the sun. It's well established. The fourth aspect of the gravitational field is the gravitational wave, the analog of the electromagnetic wave. And general relativity says that this wave should propagate with identically the same speed as an electromagnetic wave in vacuum, the speed of light. Like an electromagnetic wave, what it should do is produce forces perpendicular to its direction of propagation—"transverse forces." Whereas an electromagnetic wave pushes a charged particle back and forth in a transverse direction, the gravitational wave should push masses back and forth transversely. At any given location, the transverse acceleration will be the same for all masses; this is demanded by the Principle of Equivalence, just as Equivalence demands that the earth's gravitational acceleration be the same for all masses. This means that you can't detect a gravitational wave by localized measurements: the same acceleration for all masses means no acceleration of one object relative to another, which means no means of measuring the acceleration. However, there is an inhomogeneity in the acceleration, in that if you look at two separated masses, separated transversely to the direction of propagation of the wave, the wave should produce a relative acceleration proportional to their separation. Thus, the manifestation of the wave is a transverse strain in space—a transverse relative displacement of masses that is proportional to their separation.

That is your tidal force.

That is your tidal force, but it's a strictly transverse tidal force, and it's a dynamical transverse tidal force; it's one that propagates at the speed of light. It shares with the ordinary tidal force—the force by which the moon and sun produce the tides on the earth—this property of being proportional to the separation between measurement points. But in other respects it's rather different from the ordinary tidal force: different in its propagation, and different in being strictly transverse. Now, every kind of propagating wave in fundamental physics, one that's propagating at the speed of light, is known to be associated with some sort of zero rest mass particle. Electromagnetic waves are associated with photons; gravitational waves are associated with gravitons. Electromagnetic waves are created, in fact, by a superposition of a large number of photons, all in the same quantum state; the photons carry the wave; they are the wave. Gravitational waves similarly are created by a large number of gravitons, all in the same quantum state. The gravitons carry the gravitational wave; they are the wave. By looking at the

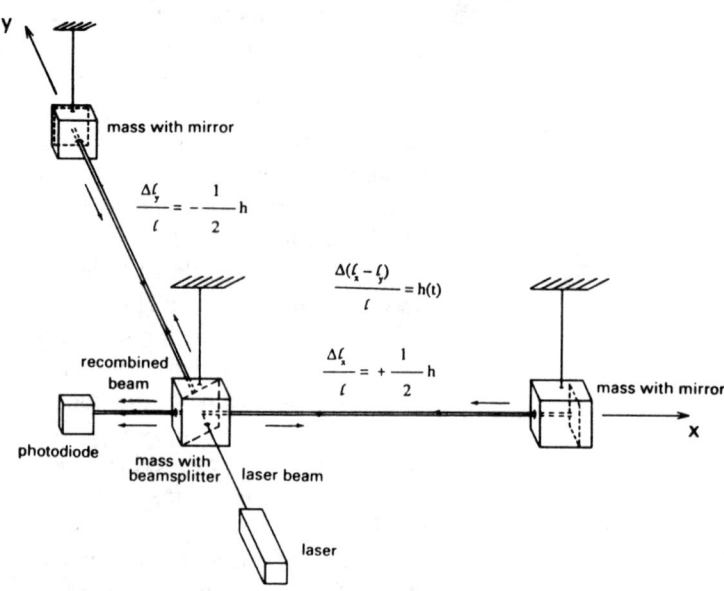

Figure 3. Schematic diagram of a laser interferometer gravity-wave detector. Three pendulum masses are hung on corners of an L. The quantity h is the dimensionless gravity-wave field (strain in space) impinging on the detector; $\Delta l_x / l$ is the fractional change in the leg length produced by that field along the x-direction; and $\Delta l_y / l$ is the fractional change along the y-direction.

classical features of the wave field, you can deduce properties of the underlying quantum mechanical particle. When you look at an electromagnetic wave, the very fact that it propagates at the speed of light enables you to deduce that the underlying particle has zero rest mass. If you look at the polarization properties of the electromagnetic wave, you notice that at any given moment of time, the acceleration it's producing on a charged particle is a vector acceleration, a directional acceleration orthogonal to the propagation direction. That acceleration pattern is something which, if rotated about the propagation direction by 360 degrees, returns to its original form. This is a property, of course, of any vector: you rotate it 360 degrees and bring it back to where it was. Now, there is a theorem in fundamental physics which says that this return angle, the angle through which you have to rotate the transverse acceleration pattern in order to get back to where you started, is equal to 360 degrees divided by the spin of the particle that carries the wave. And from the fact that the electromagnetic return angle is 360 degrees, we deduce that the photon has spin 1. Well, for gravitational waves the return angle is 180 degrees. We know that because it's a transverse tidal force, it's something that stretches things in opposite directions—it stretches things in the east/west direction and squeezes them in the north/south direction if the wave propagates vertically through the earth (Figure 1). That's a pattern which, at any moment of time, if rotated about a vertical direction, returns after 180 degrees to its original form. The return angle is 180 degrees. Divide that return angle into 360 degrees, and you will get 2, which has to be the spin of the graviton. So one is faced with the exciting prospect that if one can measure the propagation speed of gravitational waves, and find that it's precisely the same as the speed of light in accord with general relativity's prediction, one will thereby conclude that the graviton has zero rest mass. And if one measures the polarization properties of the wave and discovers that they indeed have a return angle of 180 degrees, as predicted by general relativity, one will thereby deduce that the graviton has spin 2. It would be very exciting to actually see that experimentally. And, of course, there's always the tantalizing possibility that the answers may not come out the way they're supposed to—and we would thereby have to conclude that the graviton does not have the properties that Einstein predicts.

With the planned laser interferometer gravitational wave observatory, you're hoping to first of all detect gravitational waves, and secondly, to find some of the answers to questions that you just posed.

Right. Except you shouldn't say that I am hoping to do so, because I'm just a flaky theorist who talks to and encourages experimenters who have the real talent and the real ability to build such an observatory. People like Rainer Weiss and his colleagues at MIT and Ronald Drever's group at Caltech, Jim Hough's group in Glasgow, Scotland, the group that was started by Heinz Billing in Garching, West Germany, and the group of Alain Brillet, Paris— these five groups have some of the most talented experimental physicists in the world, and they are all pushing toward building LIGOs—laser interferometer gravity wave observatories, which would make possible experiments of this sort, opening the new gravitational wave window onto the universe and extending fundamental physics.

Can you give me a brief background on the LIGO?

The American laser interferometer gravity wave observatory, the LIGO, is a proposed large-scale facility for which small proto-types have been constructed and operated since the mid-1970s. The kind of detectors we're talking about are very different from the original bar detectors that Weber pioneered. They are made up of three masses that hang as pendula from overhead supports. One of the masses is at the corner of an L, and the other two are at the ends of the L. When the gravitational wave comes by, if its frequency is high compared to the one Hertz swinging frequency of the pendula, it should push the masses back and forth relative to each other just as though they weren't hung. Because of its polarization properties as predicted by general relativity, the two masses along one leg of the L should be pushed together while the two on the other leg of the L are pushed apart. And then in the second half cycle the wave should push the two on the second leg together while those on the first leg are pushed apart. The idea, then, is to use laser interferometry to monitor these relative motions of the masses by monitoring the difference in the lengths of the two legs of the L. This is done by shining a laser beam onto a beamsplitter that rides on the corner mass, and splits the beam in half, directing the two beams down the two arms to mirrors on the end masses. The end mirrors reflect the beams back to the beamsplitter, which recombines them. A change in the relative arm lengths causes a change in the relative phase of the recombining light, and there is a change in the intensity of the recombined light coming out each side of the beamsplitter. That at least is a flaky theorist's version of the experiment—of course the real experiment is far more sophisticated than that. It involves either operating the legs as delay lines, with the beams bouncing back and forth many times, making many discrete spots on the mirrors, or operating each leg as a Fabry-Perot interferometer with the beam, in effect, bouncing back and forth many times but making only one discrete spot on each mirror. And it involves dozens of feedback loops to keep the interferometer "locked" and dozens of tricks to avoid dozens of sources of noise.

I read that the LIGO is going to span the continent. What does that involve?

With any kind of a gravity wave detector, you face the problem that there are frequent, unexpected, poorly understood, sudden vibrations or jerks in the apparatus that you can't distinguish locally from a gravity wave. For example, there may be a sudden, tiny strain release in one of the wires from which one of the masses is hanging, which might cause the mass to jerk slightly. You don't know whether this is due to a gravity wave when you see it, or whether it's due to a sudden strain release in the wire. The only way to clearly determine whether something that is seen is a gravity wave is to operate two or more such instruments simultaneously at distantly located sites, and then cross-correlate the output from them. And so the proposal in the United States is to build one such LIGO consisting of two sites: an L with four-kilometer legs on the northern edge of Edwards Air Force Base in Southern California, and another L in Columbia, Maine, in one of the world's largest blueberry patches. These two sites would be built and operated by a consortium of Caltech and MIT physicists, and would be operated in coincidence, with the data cross-correlated in order to search for and study the gravitational waves.

What sensitivity do you think the experimenters can get from such a device?

The sensitivity range in which I, as a theorist, expect is for the strongest waves to lie somewhere between 10^{-20} and 10^{-22} for the strain, the fractional change in leg length produced by the waves. If I had to pick a number out of a hat, I'd say 10^{-21}. However, we don't know for sure how strong the waves are, by any means. The reason is that all our present knowledge is based on electromagnetic studies of the universe, and those electromagnetic studies probe the wrong things, like the surfaces of collapsing stars, not the cores, for example. So there's an enormous unknown about the waves. Fortunately, we are fairly confident of one source: the spiraling together and coalescence of two neutron stars in a binary system in a distant galaxy. We are fairly sure that, if you want to see such an event, say, several times per year, you will have to look out to a distance of 300 million light years—give or take a factor of two—and we know that the waves from that distance would produce a strain of about 4×10^{-22} on earth. The detectors that are planned for the full scale LIGO should have sensitivities, initially, of a least 10^{-20} and probably better. The LIGO will be constructed to house many generations of ever-improving detectors. Remarkably, the experimenters know how to build, with essentially existing technology, detectors that are as good as 10^{-23}. However, pushing the actual sensitivities from 10^{-20} down toward 10^{-23} is going to be very, very difficult, because the detectors involve huge numbers of servos; they involve making an exceedingly complex system work exquisitely well. So what I anticipate is that, when the first detectors go on line in the early 1990s, they will have sensitivities a little better than 10^{-20}; and the experimenters will work very hard to improve the sensitivities, building one generation after another of detectors in the big LIGO vacuum system, pushing their way down to the 10^{-20} level. Somewhere along the way, perhaps at 10^{-23}, almost certainly before 10^{-23}, they will hit waves and start doing exciting astronomy.

It seems to me that there would be a lot of thermal effects in all of this, and the equipment would have to be made superconducting.

It turns out not. The present generation of bar detectors are cooled to liquid helium temperatures to circumvent thermal noise. However, laser-interferometer detectors have the big advantage that the strength of the signal is proportional to the length of the legs. We're talking about laser detectors that have leg lengths of four kilometers, as compared to bars with lengths of, say, two meters. That's a ratio of two thousand, which means that for a given level of sensitivity, thermal problems in bars are a factor of 2,000 worse in amplitude, or a factor of 4,000,000 worse in energy for a given sensitivity than in the laser systems. This is why the bars must be cooled to keep thermal effects under control, and why it does not look like one will need to cool the laser systems, at least not in this century. I would be extremely surprised if one had to go below room temperature in order to start doing interesting astronomy with a laser system.

What about the cost, and when is your projected completion date for such an observatory?

The estimated cost is about $60 million, in 1984 dollars, for the planned system in the United States. That buys you two L-shaped vacuum facilities in which a variety of gravity wave instruments can be built and operated over a period of, say, 20 years or more. In Britain, where money is much tighter, they initially plan to build one facility with one kilometer long legs and then upgrade it to three kilometers. In Germany they're pushing initially for one facility with three kilometer legs. The British and Germans will cross-correlate their data with each other and with the Americans. In each of these countries the hope is that work can be completed and the first gravity wave experiments can start occurring somewhere between 1991 and 1994. When they begin depends largely on the politics of getting these large amounts of money.

Do you have anything else you'd like to add?

Only to re-emphasize that I'm just a theorist. I'm not doing the real experiments. The design and construction of the detectors, and the rich technological fallout from that effort, come from the clever hands and fertile brains of people like Drever, Weiss, Hough, Billing, and Brillet, the real experimenters. They're the people who have the talent. If I had any experimental talent, I wouldn't be doing the theoretical things that I do; I would be in the laboratory working on these detectors, because I think that's where the greatest excitement is going to be over the coming years. I wish I had that talent.

Update: The perceived completion dates would be around 1994 and 1995. The sites themselves will be chosen by the National Science Foundation, and may not be at Edwards Air Force Base and the blueberry field in Maine. The cost is expected to be $45 million a year for four years.

Clyde W. Tombaugh

The Discovery of Pluto

Some Generally Unknown Aspects of the Story

Clyde W. Tombaugh holds the distinction of being the only living man to have discovered a planet. He is Emeritus Professor of Astronomy at New Mexico State University.

Clyde Tombaugh in 1931. He is shown at the eyepiece of Lowell Observatory's 7 1/3-inch guide telescope of the 13-inch astrograph, which he used in his discovery of Pluto. (Lowell Observatory photograph.)

How I Came to Flagstaff

I was born on a farm near Streator, Illinois, on 4 February 1906. In 1922, our family moved to a large wheat farm in western Kansas, where I had clearer skies. I used a Sears Roebuck 2 1/4 inch telescope hundreds of times on the Moon and planets. Also, I observed a transit of Mercury, star clusters, and nebulae, including the Crab Nebula.

I wanted a more powerful telescope so I started grinding mirrors. The third one was a 9-inch of excellent quality, which yielded sharp images under a high magnifying power of 400 diameters. In the fall of 1928, I made many sketches of Jupiter and Mars at the eyepiece of the 9-inch. I made exact copies of them and sent them to the Lowell Observatory at Flagstaff, Arizona, for their comment. Unknown to me at that time, Vesto M. Slipher was looking for a good amateur astronomer who might be trained to take long-exposure photographs with their new 13-inch astrograph.

I could not have sent in my drawings at a better time. They may have compared the detailed markings on my drawings with their current photographs of the planets. Anyway, from comments made in V.M. Slipher's letters to the Trustee and others, they were impressed with my work. V. M. Slipher promptly answered my letter, asking questions about my schooling, interests, and physical health. With this, I suspected more than just polite interest. I immediately answered his letter. Within a week, I received another letter from Slipher, asking if I would be interested in operating a new photographic telescope in a cold, unheated dome throughout the night. If so, would I be interested in coming to Flagstaff on a

3-month trial basis? I could not have been more eager to accept such a proposition.

Another letter arrived suggesting that I come to Flagstaff about the middle of January 1929. So I made ready for travel to Arizona, packing my math, physics, and astronomy books into my heavy suitcase. I boarded the Santa Fe train at Larned, Kansas, with not enough money in my wallet for a return ticket. After 28 hours in a chair car, I arrived at Flagstaff about 1:00 pm, 15 January. V. M. Slipher was there at the depot to meet me and took me up Mars Hill to the Observatory Administration Building, where he introduced me to C. O. Lampland and Mrs. Fox (the secretary). Then, he took me upstairs and assigned a bedroom to me. At 5:00 pm, I rode downtown with Mrs. Fox, and found a cafe to eat my first dinner in Flagstaff.

Getting Started

The next day, I was taken out to the new 13-inch telescope dome. Stanley Sykes—the machinist and instrument maker—and his son, Guy, were installing some of the hand controls on the equatorial mounting. The front end of the tube was open because the objective lens had not yet arrived from the Alvan Clark firm in Massachusetts.

Completion of the telescope was behind schedule, so I was assigned various odd jobs—shoveling snow, stoking the large furnace with pine logs, and showing visitors astronomical pictures and the long 24-inch refractor.

Finally, the 13-inch Cooke Triplet objective arrived on 11 February 1929. I watched V. M. Slipher and Sykes open the box, and there was the beautiful 13-inch jewel. There was a sense of great relief that the glass survived the long shipment journey without any damage. The following day, the 13-inch objective, in its cell, was bolted onto the front end of the tube.

During the following nights, I assisted V. M. Slipher in the dome with the new 13-inch. The first night involved making a focus test plate. Slipher adjusted the objective at various focal settings, the images being separated by small, progressive changes in the direction the telescope was pointing, while I recorded the numbers. After the plate was developed, Slipher selected the best focus and adjusted the objective to that position and locked it in.

The first few exposures were on 11 by 14 inch plates. The field was so good that Slipher decided to use 14 by 17 inch plates. To get the best possible image on plates that large, they had to be bent into a slightly concave form. Slipher had three special plate holders

Within a week, I received another letter from Slipher, asking if I would be interested in operating a new photographic telescope in a cold, unheated dome throughout the night. If so, would I be interested in coming to Flagstaff on a 3-month trial basis? I could not have been more eager to accept such a proposition.

made in the shop. Small strips of brass with slightly convex curves were fastened on the sides and ends of each plate holder for the plates to rest on for bending. There were thumb screws on the back for each corner and one for the center. The plates had to be bent with a depth of about 4 millimeters concave toward the objective lens to conform to the curvature of the focal plane. Slipher had a testing table made with a long amplifying arm to measure accurately the curvature of each plate in the plate holder before it went to the telescope. All parts of the plate had to be within 1/200th of an inch of the focal surface of our instrument!

The smallest star images were about 1/30th of a millimeter in diameter. If the star images are not exactly the same size and shape on both plates of a pair, the plates are "unblinkable" for a thorough examination with the blink comparator.[1] Also, the duplicate plate had to be made at the same angle above the horizon in order to match the slight distortions by atmospheric refraction for portions of the plate several degrees from the guide star.

To take exposures as long as an hour, we used a brighter guide star which we kept centered in our eyepiece. V. M. Slipher soon had me do the guiding on the plate exposures. After watching my performance for a few nights, he finally said to me "You are doing all right; you are on your own," and he stopped coming out to the dome.

Disappointment and a New Task

After I had obtained the three pairs of large plates spanning the constellation of Gemini, V. M. and E. C. Slipher started blinking them in an effort to find Planet X quickly. They took turns working at it for about a week. Two of the plate regions had about 300,000 star images each. To pick one tiny image that shifted position among the myriad star images is an awesome task. I have often wondered if the Sliphers took into account that the plates were taken only a little past the western "stationary point,"[2] which would

---

1. In a blink comparator, the same parts of two photographic plates taken some time apart are alternated (blinked) in a viewer. If the plates are aligned, stars that have not moved in the time between the photographs will look steady to the viewer; objects that have moved will appear to "blink" (shift position against the star background) as one plate and then the other are viewed in rapid succession. Such an instrument is tremendously useful for making out moving objects among a huge number of stationary images. It is also useful in finding variable stars.—*Mercury* Ed.

2. As Earth catches up with and passes a slower-moving outer planet, there are two times when the outer planet appears (from Earth) to stop and change the direction of its motion against the background of the stars. At such a time, the planet is said to be at a "stationary point."—*Mercury* Ed.

drastically affect the rate and direction of the apparent shift in position. In any case, the Sliphers missed seeing the faint images of Pluto.

I happened to be present when V. M. Slipher removed the last of the three pairs of Gemini plates from the blink-microscope comparator. "Did you find Planet X?" I asked. Slowly and sadly he said, "No. We didn't find anything." I saw then that he was terribly disappointed. I was disappointed, too, and in my own mind, wrote Gemini off for containing Planet X.

Recalling this incident in later years, I realized that Slipher was in a serious predicament. They all had done a superb job in designing and constructing the excellent 13-inch astrograph. Prior to 1927 or so, the Sliphers and Lampland had explained to the observatory Trustee that a larger and more powerful instrument was essential to finding Lowell's Planet X. Now they had the telescope. Slipher had stuck his neck out to Putnam, and Putnam had stuck his neck out to his Uncle Lawrence Lowell, who had furnished the money to build the telescope. It was imperative that they find Planet X. Slipher must then have considered that Planet X was somewhere else in the Zodiac, because he instructed me to continue photographing the Zodiac eastward. He said nothing about photographing at the opposition point, nor about taking three plates of each region, points we shall come back to. By accident, some of the regions had third plates, but were a little underexposed since they were cut off by haze moving in. I pushed the photographing through Cancer, Leo, Virgo, Libra, reaching Scorpius and western Sagittarius by the end of the June lunation. I had taken nearly 100 plates.

Except for a little blinking by Mrs. Fox (the secretary), no other blinking was done by my senior colleagues. This worried me, but I felt I dared not ask why. Who would do the blinking, and when? After the blinking of Gemini in May, I think Slipher was demoralized. Of course, he wanted to be the one who found Planet X. Undoubtedly, he realized that the task greatly exceeded the amount of time he could devote to it. My three senior colleagues were busy and in arrears of their regular observatory programs.

At the end of the June lunation, Dr. Slipher came to my office and said that they wanted me to start blinking the pairs of plates! I shuddered at the thought of having to examine all those hundreds of thousands of star-images. Also, I was nearly exhausted from loss of sleep from the heavy night runs at the telescope.

What did this new assignment mean? I thought, "Are they throwing in the towel?" I had no college education and relatively little observatory experience. At the time, I did not realize how

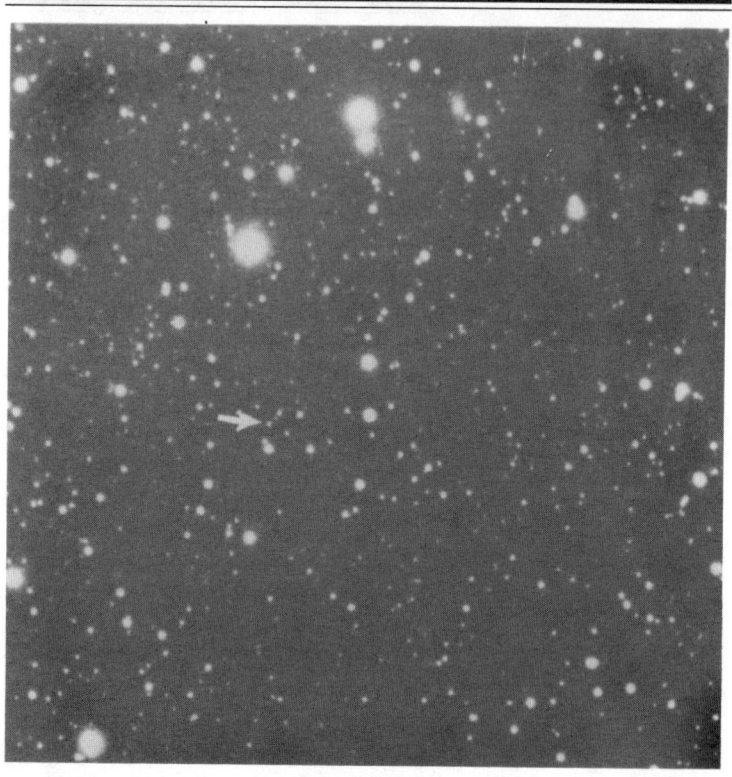

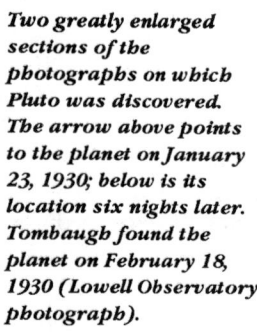

Two greatly enlarged sections of the photographs on which Pluto was discovered. The arrow above points to the planet on January 23, 1930; below is its location six nights later. Tombaugh found the planet on February 18, 1930 (Lowell Observatory photograph).

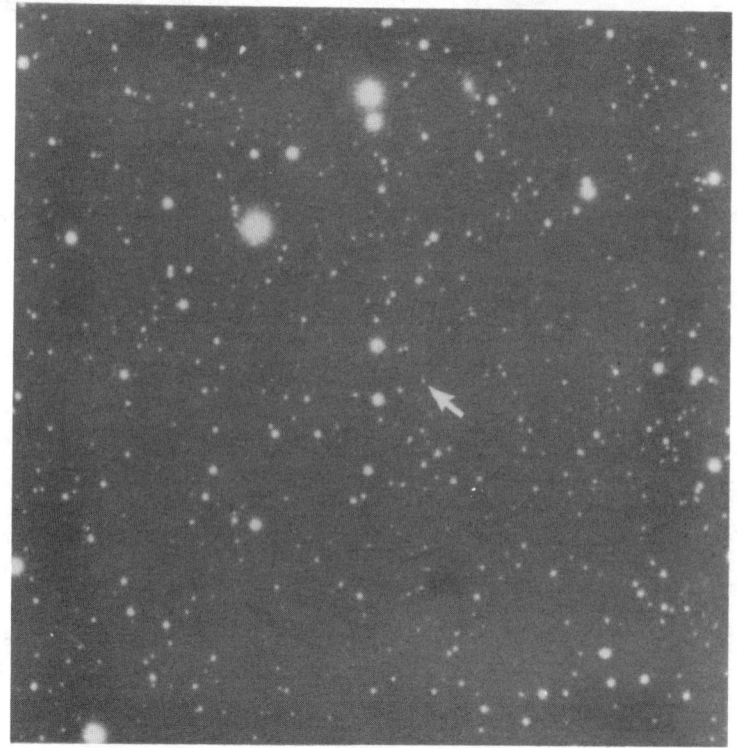

desperate the situation was. Perhaps my vigorous and vigilant dedication to making excellent plates raised their confidence in me?

Well, I blinked two pairs of plates, and encountered lots of variable stars and asteroids. How was I to distinguish Planet X from all of those asteroids, which also moved? I found the blinking to be very tedious. Would I have to go on blinking a hundred or more pairs of plates before I would find Planet X? The task seemed overwhelmingly formidable, and I became discouraged.

A Better Technique

The Rocky Mountain summer monsoon was beginning to move into the Flagstaff area in early July. The rainy season continued into early September. Most of the nights were very cloudy. Dr. Slipher suggested that I take my vacation at this time. I boarded the Santa Fe train and headed home to Kansas.

When I returned three weeks later, it was with some sense of dread. Since the blinking would require so many hundreds of hours of tedious work, my thought was to do the scanning so thoroughly that a given region could then be "written off" for containing Planet X. I did a lot of hard thinking on how to eliminate confusion with the hundreds of asteroids that I would encounter. Then I remembered the motions of Mars, Jupiter, and Saturn that I had observed in my youth. I realized that there was a definitive pattern of the planets' motions as seen from Earth.

Wishing to study the apparent motions in a more quantitative way, I plotted the apparent motions of Uranus and Neptune through two years from their positions given in the American Ephemeris and Nautical Almanac. I saw the solution to my problem—photograph the regions of the Zodiac only when the region I was searching was at opposition (on the opposite side of the Earth from the Sun). Since the Earth moves much more quickly around the Sun that an outer planet does, it is just at this point that the outer planets show the maximum retrograde motion—appearing to move "backwards" in our sky as the Earth "leaves them behind." When planets are at this opposition point as seen from Earth, all their relative motion is perpendicular to our line of sight (across the sky). In this case, there is a simple relationship between a planet's or asteroid's distance and its apparent motion—the more distant the object, the less it will appear to move.

What this means in practice is that when the region I was photographing was at opposition, the asteroids would move about seven millimeters per day on the plates, while Pluto turned out to move only half a millimeter per day. Clearly, if the daily shift of a "blinking object" was less than even that of Neptune, the object would definitely be trans-Neptunian.

Then I went to see V. M. Slipher in his office about what I had discovered about planet search strategy. I said that most of those 100 plates would have to be done over since they were taken too far from opposition. Well, he agreed.

It has always baffled me why no one discussed this with me earlier. Later, it became evident to me that they were so obsessed with a "quick find" attitude that their observational procedures were defeating them. Of course, one of the factors was that they were looking for a 13th magnitude planet, as Lowell had predicted. Apparently, they were scanning the plates much too fast. One has to actually see about 12 times more star images to detect a 15th magnitude planet.

It was not until I read the manuscript of William Hoyt's book *Planets X and Pluto* in 1979 that I learned that Lowell had recommended taking the plates at opposition, a fact Hoyt had extracted from some of the early archives. Apparently, the early observers did not take Lowell's recommendations seriously and thus I had to re-discover this strategy for myself.

The Discovery

The rainy summer monsoon ceased about the middle of September. With clearing skies, I started photographing in the Zodiac belt again. Since the season was now near the Autumnal Equinox, I started taking plates in the constellations of Aquarius and Pisces. With each succeeding lunation, I marched 30 degrees eastward to keep up with the eastward moving opposition point. These were non-Milky Way regions where the sample star counts indicated only about 50,000 star images per plate. I could blink a pair of plates thoroughly in three days of hard work. These regions were loaded with those beautiful spiral nebulae. I started making counts of them for each of the ten panel sections on each pair of plates. Their distribution was quite uneven, containing from about 100 to over 300 galaxies per plate.

Before beginning the blinking of a new pair, I would note the interval in dates between the plates of a pair, in order to anticipate the shift of Planet X to look for. The scale on the plates was almost exactly 2 arc minutes per millimeter. There were always a dozen or more asteroids on each pair of plates. Their images were short trails caused by their motion during the one hour exposure on each plate. In this series of plates, I encountered Uranus, which had the amount of shift that I expected. Now I felt that I had the planet search under full control for a thorough and systematic search.

In late November and early December, I was photographing in the constellation of Taurus. With each passing lunation, I was getting closer and closer to Gemini. The night work ceased about December 8th because of the moon. From then to Christmas, I spent all my time blinking the plates in western Taurus. After Christmas, the night work of photographing was resumed for two weeks. For nearly three weeks in January, I struggled with blinking other pairs of plates in Taurus. The number of stars greatly increased as I approached the Milky Way, which slowed down the blinking coverage. I realized then that the Sliphers had blinked those 1929 plates of Gemini much too fast, so I decided to rephotograph Gemini, this time near opposition.

On 21 January, I set the 13-inch telescope on Delta Geminorum (a 3rd magnitude guide star). It was an extremely clear night. About 10 minutes after I opened the shutter, a fierce northeast wind came up. As the intense gusts swept up Mars Hill, the star image expanded in size to several angular diameters of Jupiter, becoming so diluted that I could see nothing of it. I was frantic, I couldn't see anything to guide on. Fortunately, the driving clock was working well. After the gusts, the image of Delta became visible again, pulsing in size and with agitated motion. Then another gust and invisible again.

To those of you with experience in guiding telescopes during long exposures, imagine seeing so bad that a 3rd magnitude star disappears in a 7 1/3 inch guide telescope on a crystal clear night! In all of the following years, I have never again experienced such bad seeing. As it turned out, this was the plate that recorded the swollen image of Pluto. The Delta Geminorum region was still jinxed. With this kind of luck, I should have been suspicious that fate was trying to hide a planet image from me.

At the end of the hour's exposure, I closed the dome shutters and took the other unused loaded plate holders back to the Administration Building and called it quits for that night.

On the nights of 23 and 29 January, I re-photographed the Delta Geminorum region. I would have preferred a two or three day interval, but I had other plate regions to get started. Also, I may have had a cloudy night or two, I don't remember. Anyway, it was nearly always a lively scramble to keep up with the eastward-moving opposition point.

During the "light of the moon" period in the second and third weeks in February, 1930, I blinked the easternmost pair of plates of Taurus, which is right in the Milky Way. Each plate of the pair contained about 400,000 star images! The blinking was awfully tedious. I had had enough of the Milky Way for the time being, and decided to postpone the two, very star-rich, regions of western Gemini.

Since the new lunation of night work was about to start, I decided on the Delta Geminorum pair. By four o'clock in the afternoon of 18 February, I had blinked one-fourth of the pair. It was still moderately rich in stars, containing about 1000 per square degree. Upon starting a new strip to the east of the guide star, Delta, as I turned my attention to the second field over, I suddenly spied a 15th magnitude image appearing and disappearing. Glancing to one side of it I found another 15th magnitude image doing the same thing. "That's it!" I exclaimed to myself. I saw the images almost instantly in a 1 by 2 cm field containing about 300 stars. (I always used a rectangular diaphragm over the field lens of the comparator eyepiece so that I wouldn't have to blink the areas near the edge of a circular field twice.)

I stopped the automatic blinking, and flipped the views back and forth with the finger lever, intently studying the images and shift in position. I had never encountered a planet suspect so promising during all of the fall months blinking. With a plastic ruler, I measured the shift to be 3 1/2 millimeters. Oh, I thought, I had better look at my watch, this could be an historic moment. It was within about 2 minutes of 4 pm (MST).

Ah! The object ought to be about one millimeter to the east of the 23 January position on that miserable 21 January plate. So I removed one of the plates and placed the 21 January plate on the comparator. Sure enough, there was the swollen image of Planet X (Pluto) exactly where it should be. The amount of shift indicated that the object was far beyond the orbit of Neptune. I was now nearly 100 percent sure. I got out the three 8 by 10 inch plates taken with the 5- inch Cogshall camera, along with the 13-inch plates. The object's images might just be recorded, barely. Inspecting the plates with a hand magnifier, identifying the configurations of faint stars, there they were on all three plates, exactly in the same corresponding positions. I was so excited by this time that I could hardly hold the magnifier steady enough. For three quarters of an hour, I was the only person in the world who knew exactly the position of Planet X.

Confirmation

C. O. Lampland was sitting at his desk across the hall. He heard the clicking noise of the blink comparator suddenly stop and then a long silence. He thus suspected that I had run onto something and was nearly dying of suspense (he told me later). I called him in to view the new planet images and I explained that the shift was right for a trans-Neptunian planet.

Then I went down the hallway to V. M. Slipher's office. I paused a few moments to get myself into as nonchalant a mood as possible. His door was open. Slipher was sitting at his desk, working with some papers. I boldly strode into his office. "Dr. Slipher, I have found your Planet X," I said. He rose up from his chair, as if propelled by a spring, with a facial expression of excitement but reservation in his voice. I don't remember exactly what he said. Then I said, "I can show you the evidence." He was on his way to the blink comparator room so quickly, I had to step lively to keep up with him.

Lampland surrendered the comparator to Slipher, commenting, "It looks pretty good." The air was tense with excitement as I interchanged the January 21st check plate for his inspection of the confirmation. Finally, Dr. Slipher said to me, "Rephotograph the region as soon as possible." I looked out of the window at the sky. It was pretty much overcast. "Doesn't look very promising for tonight," I said, "but I will sit up for it through the night." Lastly, Dr. Slipher charged, "Don't tell anyone about the discovery. It could be very hot news. We need to keep it secret for a few weeks to study the object." It was nearly six o'clock before we left the blink comparator room. Generally the staff members went home about 5 o'clock.

I drove downtown to eat my dinner in a cafe and perform my usual duty of picking up the observatory mail at the Post Office. After dinner, I noted that the sky was heavily overcast. I was extremely excited and had to calm myself. So I went to the Orpheum Theatre to see Gary Cooper in "The Virginian." I can never forget that night. After the gun-fight, my knees were shaking more than ever.

Jerry Nelson, Fritz Merkle, Frederic Chaffee

Telescopes for the 1990s

Jerry Nelson (top left) is Project Scientist for the Keck Observatory, a ground-based 10-meter telescope with an f/1.75 segmented primary mirror, a joint project of the University of California and California Institute of Technology. He received the PhD in physics from the University of California at Berkeley in 1972.

Fritz Merkle (bottom left) is head of the Optics Group, Technology Division, at the European Southern Observatory (ESO). He received the PhD from the University of Heidelberg in image processing, pattern recognition, and adaptive optics. His main technical involvement is in the development of active and adaptive optics for astronomical telescopes and spatial interferometry, in particular, for the Very Large Telescope (VLT).

Frederic Chaffee (right) is Director of the Multiple Mirror Telescope (MMT) Observatory, operated jointly by the Smithsonian Institution and the University of Arizona. He received his PhD in astronomy from the University of Arizona in 1968. His research interests have evolved from high resolution spectroscopy of stars to the study of the interstellar medium to, most recently, spectroscopy of quasars.

The three were interviewed by Frederick Su at SPIE's Astronomy symposium in Tucson, Arizona, February 1990.

Fred, what new things have been done with the MMT?

Chaffee: In its current configuration, the most important thing that has happened has been the implementation of a phasing capability that allows us to keep all six telescopes phased in the near infrared for arbitrary periods of time. For us, that was the last big technological question we wanted to answer since we last spoke. That has in fact been implemented.

I heard you talk also of a seventh mirror that can go down to .125 arcsecond.

Chaffee: Right. We have replaced one of the six 1.8-meter mirrors with an extremely high quality one. The original mirrors were capable of producing images of roughly .5 arcsecond. We realized very soon after the MMT was commissioned in 1979 that that was not adequate to take advantage of the superb conditions at the site. So we began a project in 1984 to figure a seventh primary mirror to much higher precision—capable of producing 1/8 arcsecond images. That mirror was completed in late 1987 and installed in the telescope in 1988. Our initial intent had been to refigure the original six mirrors to that precision. However, in the meantime Roger Angel has developed the technology to produce rigid lightweight mirrors up to eight meters in diameter, so we have chosen the path of replacing the original six 1.8 meter primaries with the largest single mirror that can be fit into the existing telescope structure.

What is the effective aperture of the MMT?

Chaffee: Currently, with the six original mirrors, the effective aperture is 4.5 meters. Using the technology developed by Roger Angel at the Steward Observatory Mirror Laboratory, we expect to install a single 6.5 meter primary in the MMT by 1994.

Even for moderately bright IR sources, the background sky produces a million photons for every single photon from the source. As an optical astronomer, I'm not used to thinking in those terms. Such a low contrast pushes you to limits that optical astronomers rarely have to think about. Infrared astronomers have to do very clever things in order to modulate out this sky noise in some way.

The Multiple Mirror Telescope Observatory at dusk. Photograph by Diane Nilson.

That leads us to you, Jerry. What is the aperture of the Keck telescope and how does it differ from the MMT?

Nelson: Ours is a 10-meter diameter telescope. So it is substantially larger than the MMT. It is similar to the MMT in that it has multiple mirrors. Our primary mirror consists of 36 hexagonal segments. But, in concept, it is much more like a conventional telescope in that our 36 segments effectively form a single continuous primary mirror that has a single focus. The MMT, in contrast, has six mirrors; each of them has its own focus, and through some additional optics, that's combined to form a single focus.

Now I've heard the words "active optics" thrown around. I think we have two definitions of active optics here. We have the MMT and yours, Jerry, with all those actuators.

Nelson: Our active optics is required to keep the mirrors in their correct positions and orientations, relative to each other. We have a set of sensors that measures where those mirrors are relative to their neighbors. A mathematical control system, run through a computer, then directs the set of actuators that move the segments. It is a closed loop system. We sense where we are, we move to where we want to be. That is an active system.

Inside the dome of the Keck Telescope. © 1990 Roger Ressmeyer— Starlight. Photo courtesy of the California Association for Research.

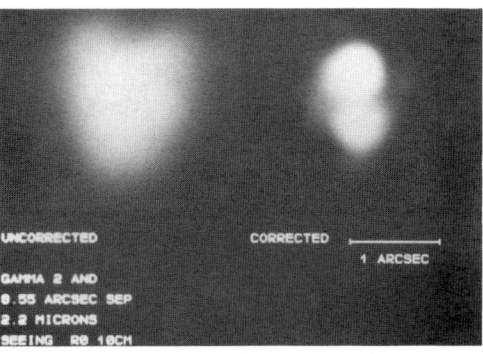

First diffraction-limited image of a binary star obtained with the ESO adaptive optics prototype system in the near infrared. The image shows Gamma 2 And in the K-band (integration time 4 seconds). The separation of the binary is 0.55 arcseconds. Left: The uncorrected image. Right: The image with adaptive optics correction. Photo courtesy of European Southern Observatory.

Chaffee: Conceptually, the two telescopes are actually quite similar. We move secondary mirrors instead of segments of the primary mirror to produce a single image. In recent years, the community has evolved a consistent vocabulary in dealing with two related, but distinct, ways to improve telescope images. The two terms used are "active optics" and "adaptive optics." Active optics has come to mean any means by which the alignment or surface of the telescope optics can be adjusted to compensate for low frequency changes in the telescope structure or the primary mirror surface. By this definition, the MMT, the Keck telescope, and the NTT all employ active optics. The MMT compensates for gravity deflections in the optics support structure by tilting the secondary mirrors. The Keck will compensate for relative gravity deflections of the primary mirror segments by independently controlling each segment. And the NTT compensates for gravitational distortion of the thin meniscus primary mirror by sensing the quality of the mirror surface and flexing the mirror locally by means of any of the 78 controllable pistons that support the meniscus. For all three telescopes, these corrections are made rather slowly—in the range 0.01 to 1 Hz or so. Adaptive optics has come to mean any means by which telescope optics can be adjusted to compensate for distortions caused by the earth's atmosphere in the incoming wavefront. To realize significant gains over active optics techniques, corrections must be made spatially and temporally and at high frequencies, usually by means of a small, deformable mirror onto which the telescope pupil can be re-imaged. Such corrections must be made at spatial scales smaller than about 30 cm on the incoming wavefront and at frequencies up to 250 Hz. If such corrections can be made, the gains are impressive. In principle, one can achieve the diffraction-limited resolution of the telescope—an achievement until recently considered possible only from above the earth's atmosphere. What we have at the MMT is an active optics system that operates at about 1 Hz.

Are you sensing at that frequency?

Chaffee: We actually can sense at video rates (30 Hz) if we have a bright enough guide star in the field. However, we issue correction commands to secondary mirrors much more slowly than that—typically at 0.1 Hz. This slow rate is satisfactory primarily because the mechanical structure of the MMT is so rigid that gravitational deflections are slow and of small amplitudes.

That brings me to you, Fritz. Tell us about your new 3.5-meter telescope, the NTT (New Technology Telescope)?

Merkle: In the NTT, we have active optics. The primary mirror is resting on 78 actively controlled actuators. The image quality is measured with a wavefront analyzer that sends corrections approximately once every 30 seconds to correct the image quality in a closed loop system. On the other hand, we have worked on a prototype system for adaptive optics, which is a much faster correction, as Fred stated. The system corrects the wavefront at approximately 100 times a second. These corrections—or in other words the aberrations that need to be corrected—are mainly due to atmospheric problems in the air column above the telescope.

What is the aperture of your prototype system?

Merkle: The prototype is a type of instrument that is mounted to a telescope. The first test that we have done is with a 1.52 meter telescope at the Observatoire de Haute Provence in France. This prototype has actually achieved the diffraction limit at 2.2 microns. In April, we will go with the same instrument to a 3.6-meter telescope at La Silla in Chile. There we expect diffraction limited observation at 3.5 microns.

The New Technology Telescope at La Silla. Photography courtesy of European Southern Observatory.

I've heard a lot of talk about 8-meter telescopes. First off, the Hubble telescope is going up soon. That is 2.4 meters in diameter. That is going to put us up above a great deal of the earth's atmosphere, so that we will have great seeing. Now, Fritz has talked about adaptive optics, which will get rid of the atmospheric distortions ... [laughter]

Nelson: It's not the same thing.

... and then all the talk about 8-meter telescopes. What are we looking for in the future for telescopes? Are we looking for less distortion, better resolving power?

Nelson: We want more light and we want it concentrated as finely as we can get it. Hence, the virtues of adaptive optics or putting a telescope up in space. Unfortunately, at least for the next decade or so, adaptive optics is not going to sharpen up optical images. It will work in the infrared, which is where Fritz has demonstrated its usefulness. But for the optical domain, I think we have a long way to go. And that is why a space telescope wins out over any ground-based telescope. It gives superior angular resolution in the visible range and, of course, in the ultraviolet, which we can't do anything with from the ground. But when you need to get a spectrum of a star, that's when 8-meter telescopes, 10-meter telescopes, 16-meter telescopes—as big as you can get—are needed to study the physics of those objects in order to see what's going on, where the light's coming from, etc.—answers that can be gleaned in detail from its spectroscopic properties.

Chaffee: The key is that imaging is not the end-all and be-all of large telescopes. By far the majority of the time on any major telescope is spent on spectroscopy.

Merkle: I think about 80 percent.

Chaffee: 80 percent is typical. As imaging capabilities improve, they will drive a larger fraction of astronomical science toward using imaging as a fundamental technique. Nevertheless, what Jerry says is right: There are many problems for which brute aperture wins. And the combination of large aperture with superb imaging is what the new generation of earth-based telescopes is aiming at. Although, I think that adaptive optics, pushed from the infrared, will become viable in the visible, I agree with Jerry that it is going to take a while. But it will happen. Then earth-based imaging will be quite competitive with that achieved by telescopes in space. As all the speckle and image processing techniques mature, we are going to see a fundamental change in the efficiency with which earth-based telescopes use that precious trickle of electromagnetic quanta.

Exterior view of the Keck Observatory, Mauna Kea, Hawaii. © California Association for Research in Astronomy.

Merkle: One can define efficiency of a telescope as a square of the ratio of the diameter of the telescope aperture over the image size. This is the only value that tells you how efficient the telescope is. It is not the diameter only.

Chaffee: So it doesn't do you much good to double your telescope aperture if, in so doing, you degrade the image quality. You might only come out even. You have to keep the image quality high. And the image quality is a direct function of the figuring and polishing of the mirror, the electronics, the turbulence above the mirrors, etc.

Nelson: This is for faint objects, which is mainly what we're interested in, where the sky flux dominates the star flux going into the detectors.

How far back will we be pushing, looking backwards in time towards the beginning of the universe, when the 10-meter Keck comes on line?

Nelson: For the same kind of objects that you can barely see now with a 5-meter telescope or with the MMT, we will be able to see things twice as far away, twice as far back in time. We will be halving the distance back to the origin of the universe. Of course, it all depends on the object. If an object is intrinsically extremely bright, then you can see it with today's telescope back a long distance in time, back to the earliest stages of the universe. If it is intrinsically faint, even with the Keck telescope, you'll be lucky to see it even at the edge of our own galaxy. It is a simple question to ask. But the answer isn't simple.

Astronomers are more and more interested in the IR these days. How do you design a telescope to look into the IR?

Nelson: You have to make the optics good enough, independent of the wavelength you're working in so that you are limited either by the intrinsic aperture diffraction of the telescope or by the earth's atmosphere. For visible light, the atmosphere almost always dictates the image quality before aperture diffraction. The space telescope, for example, will be diffraction-limited for most wavelengths. But the 10-meter Keck telescope will have superior angular resolution for all wavelengths longer than 2 μm. Also, when you get into the so-called thermal infrared, longer than about 5 μm, you have an astonishingly large amount of background radiation from the sky, the telescope structure, the optics itself; and so the sources that are astronomically interesting are incredibly faint. One wants to minimize the telescope emissivity so the instrument sees the minimum amount of this thermal background. When one is diffraction limited, one gains dramatically in sensitivity as the telescope diameter grows, like the 4th power of the diameter.

What kinds of magnitudes are we talking about? 25?

Nelson: No. I think 15th magnitude in the thermal IR.

Chaffee: What really matters is the contrast. Even for moderately bright IR sources, the background sky produces a million photons for every single photon from the source. As an optical astronomer, I'm not used to thinking in those terms. Such a low contrast pushes you to limits that optical astronomers rarely have to think about. Infrared astronomers have to do very clever things in order to modulate out this sky noise in some way. Chopping secondary mirrors, for example, has been used frequently in the near and thermal IR for this purpose. Perhaps more important for the future, however, have been the dramatic developments in IR detectors the last few years. Arrays are becoming more available. The whole IR detector technology is changing. It is interesting to try to design a telescope that may not be in full operation for another five years when the time constant for the change in IR detector technology is six months. You're trying to guess what the infrared technology will be. Three or four years from now, who knows? Over the last few years, the changes have been dramatic. In this conference, we're seeing some very conservative designs to provide for the IR capabilities of the upcoming generation of telescopes. Many of these design features may prove to be unnecessary. It is not at all clear at this point.

Merkle: But due to this very low contrast, it is important that even in the infrared the optical quality be extremely good. Otherwise, for a very faint object in a bright background, you will not detect it. Only if the energy of the light is really concentrated will you be able to detect it. So this old, let me say, tradition thing where an infrared telescope can be a poor telescope is not true anymore. An infrared telescope has to be a very good telescope in order to be efficient.

Fritz, can you tell us more about the NTT?

Merkle: The 3.5 meter NTT is a telescope with an active primary mirror. The primary mirror is a 9-inch thin, solid meniscus. In addition, the telescope has an actively adjusted secondary mirror. While the NTT is a small telescope compared to the Keck, it relies on the same principle of active control. There are 78 actuators used to keep the primary mirror in a paraboloidal shape. The frequency at which it is controlled is once every 30 seconds. We found out that these corrections need only be done every 5 minutes or at even longer intervals. The telescope itself is quite stable. The NTT serves as a prototype for the next generation of telescopes we are building at ESO, which is a VLT (Very Large Telescope), an array of four 8-meter telescopes having an area equivalent to a 16-meter telescope.

One thing I haven't heard at this conference is the protection of astronomical sites. I visit the Big Island of Hawaii often. I see a lot of development going on. Mauna Kea is one of the best observing sites in the world. I haven't heard of any concerted efforts by astronomers trying to work with local and state governments to ensure the "pristine" astronomical quality of such sites.

Nelson: There are lighting laws all over the Big Island that require facilities to have luminaires to point the light downward so that they don't go up in the sky. They are supposed to use low pressure sodium, and they are supposed to minimize the amount of nighttime illumination they use. So there are regulations that have been negotiated through the efforts of astronomers. In fact, there is an active group of astronomers working on keeping our skies dark everywhere. And they have been successful in Hawaii and in parts of California—San Diego and the San Francisco Bay Area. They don't talk about it here at the conference because it is not an exciting technological issue. People don't give papers on it. But there is a strong community, justifiably so, that supports this issue.

Is that going to be enough, though, when you have or will have scores of new subdivisions and new resorts popping up?

Nelson: People have done population demographics studies for Hawaii in support of arguing what kind of controls you are going to need. For time scales of 50 to 100 years, the claims are that the degradation will be minimal if people follow the present legislation that has been passed.

Chaffee: We will be dead when the problem becomes serious.

[laughter]

But what about future astronomers?

Chaffee: It is a real concern. The initial legislation that I am aware of was implemented in Tucson in the early 70s when Kitt Peak was starting to be greatly concerned about the growth of the Tucson metropolitan area. At that time, Art Hoag, who was then at Kitt Peak, started to work with city officials to get city lighting ordinances passed that would protect the astronomical industry. Hoag realized fairly early that by the 1990s, Kitt Peak would no longer be a viable observing site if the sky brightness continued to increase at its 60s rate. So he began a very strong campaign, and we were fortunate in Tucson to have ordinances passed in the early 70s. Now these ordinances have spread throughout Arizona to protect the skies at all observatory sites. Nonetheless, even with shielding and high pressure sodium lights and so on, one can support only so much population density before astronomers start to worry about their observatory sites. One natural result is that astronomers, in building new telescopes, must choose sites that are increasingly isolated. One example is Mt. Graham, which is 100 miles northeast of here. It has been selected as the location of a number of large telescopes because the skies are projected to remain dark for a long time. When you spend $60-100 million on a 10-meter class telescope, you must place it where it will not be compromised for at least a century.

I don't think La Silla will have a problem.

[laughter]

Merkle: Yes. La Silla is in the Atacama Desert in Chile. The next significant settlement is about 80 miles away. This problem I don't think will come up in the foreseeable future. There is a little bit of mining activity close by, but that is not a problem.

Nelson: For those of us with U.S. sites, I just don't think it is politically viable to prohibit human beings from living within a 100-mile radius around the telescope or to say you can't turn on lights at night.

I think the question is whether to have controlled growth rather than runaway rampant growth where you have no control over it.

Nelson: Even with controlled growth, I don't think it is politically viable for astronomers to control the growth. I think the only thing we can try to control is illumination growth by dictating the kind of light people can use; and low pressure sodium is socially tolerable. It is monochromatic light so you can't see what color shirt you're wearing. But it does put all the radiation in the sky at one wavelength or a couple of narrow wavelengths, so that the astronomer can typically filter it out. It's a compromise. You'd love to have the skies dark.

Merkle: What may come up in the future is the debris in space may hurt astronomical observing. And it is much more difficult to control this.

What of the future into the next millennium? You're talking about 8-meter, 10-meter, 16-meter telescopes, and I would guess many of those would be a multiple mirror configuration because of economics. Where do astronomers draw the line?

Nelson: It is substantially technology and cost that dictates whether you go to large segmented telescopes or arrays of smaller telescopes with beam-combining methods. For optical astronomy, once you are past 4-meter telescopes, there is no scientific difference between segmented telescopes or arrays of telescopes. With today's detectors and good optics, you're photon limited; you're not detector-noise limited. And so, whether you add the signals electrically or add the photons, it doesn't matter.

Merkle: It's like in high-energy physics, you can build larger and larger accelerators until you reach a certain point where you have no more money to do it.

It all comes down to funding.

Chaffee: I think that is really a major concern in all the approaches we've been discussing. The approach that Jerry's group is taking at Keck and the approach that the MMT took were all aimed at breaking an economic stranglehold. Simply scaling up existing telescopes, even if it were technically feasible, would be economically untenable. Even a 6-meter telescope, not to mention the 10-15 meter giants we're now contemplating, would have been prohibitively expensive. All of the new technology that we are exploring, starting with the MMT, moving on to Keck and then to the VLT, have a strong economic driver. We have been forced to explore exciting alternative technological approaches to producing telescopes with bigger aperture and better performance. And the burst of creativity has been impressive.

Merkle: This was especially true with the NTT. It was a way of proving, a way of showing that this 3.5-meter telescope, which is equivalent in size to the 3.6-meter at La Silla, was a viable technology and yet only cost one third that of the 3.6-meter.

And what price was that?

Merkle: It was about 25 million German marks for the NTT.

I heard Jerry talking before about the types of aberrations and the problem the Keck has with spherical aberrations.

Nelson: It is the way we fabricate our mirror segments. We

Even with controlled growth, I don't think it is politically viable for astronomers to control the growth. I think the only thing we can try to control is illumination growth by dictating the kind of light people can use; and low pressure sodium is socially tolerable.

discovered that both as a consequence of polishing errors and as a consequence of our inability to perfectly predict the warping from the cutting—because we polish our mirrors and then cut the hexagon—the end result is that we have a mirror that isn't good enough. And so we correct that by using these things we call warping harnesses, a set of leaf springs that we basically bolt onto our wiffle-tree support system on the individual mirror segments. We have 30 leaf springs on each segment. Thirty is a finite number and, in particular, when we look at the kind of aberrations within a single segment that we can and cannot correct, some aberrations are much more difficult to correct than others. The remark that you overheard me say earlier was that spherical aberration was particularly difficult to correct with only 30 leaf springs. And I was commenting on the contrast with the difficulties we were having— we could only get a factor of 3 or 4 improvement in the net image quality from that aberration—to what the NTT did with 78 actuators and a circular mirror where they had about 3 μm in spherical aberration. And as far as I know they effectively eliminated that. It went away and the residuals introduced by correcting for it were negligible. We could get rid of all the spherical aberration if we wanted to, but in the act of doing so, we would create other aberrations. That is what limits our improvements. And, frankly, I don't understand why they do so much better than we do. Two-and-a-half times as many actuators obviously helps. But I think they did much more than 2 1/2 times better than we would do with the same aberration. Maybe it is the circular versus the hexagonal.

Merkle: Exactly. Our selection of a monolithic circular mirror was a conservative approach, because we did not leave behind traditional polishing technology with such a choice. We selected a mirror which was thinner than a classical one. This led to low-frequency aberrations that could be corrected with an active support system. Jerry, with the Keck telescope, has encountered twice the number of problems. While the Keck also has an active support system, it is a completely new technology in terms of fabricating and polishing the segments.

Education

Only the educated are free.
Epictetus

Janet S. Fender

Education and the Woman Engineer and Manager

Janet S. Fender is chief of the Advanced Imaging Technology Branch and the Air Force Center of Expertise for Imaging Technology Research and Development. In this capacity at the Air Force Weapons Laboratory in Albuquerque, she is responsible for research and development of advanced optical technologies for laser systems, high-resolution imaging, and space surveillance for the Air Force. She is a Fellow of SPIE, has served two terms as an SPIE Governor, has chaired several conferences including the 1987 SPIE annual meeting, and is a longstanding member of the SPIE Education Committee. She was guest editor of a special issue of Optical Engineering on Multiple-Aperture Optical Systems (September 1988). For her work in multiple-aperture optics, Dr. Fender received the Crozier Prize for technical excellence from the American Defense Preparedness Association, and was named by Defense Secretary Caspar Weinberger as 1987 DOD Woman Scientist of the Year. In 1988 she was named America's Most Extraordinary Woman of the Year by the Coty Corporation, and in 1990 she received the New Mexico Federal Woman of the Year award. She has also served as a science advisor for the governor of New Mexico since 1986. She was interviewed by Roy Potter, SPIE Technical Consultant.

The role of education

Janet, what is your job with the Air Force Weapons Lab?

My position description reads "optical research physicist." I am a technical manager of more than 30 professional scientists and engineers who work on original in-house and contracted applied research.

What is your educational background?

I have a BS in physics with a minor in astronomy from the University of Oklahoma and a PhD from the Optical Sciences Center, University of Arizona.

After your PhD, did you go directly to the Air Force?

Yes, I began work at the Air Force Weapons Laboratory two days after my final defense. After getting my undergraduate degree I worked at Kitt Peak National Observatory. That was a very interesting job. I worked with the Solar Telescope and the 4-meter telescope there, primarily working on Fourier transform spectroscopy. I met Bob Shannon and Jim Breckinridge in the process—perhaps you know them from SPIE—and they encouraged me to pursue optics as opposed to general physics. So I jumped over to the Optical Sciences Center. I just couldn't believe what was in the catalog! There were courses in holography, diffraction, lens design, the whole gamut of optics. They all sounded more like electives—more fun than the hard-core, drudgery physics, which I didn't think was as exciting. I was just delighted.

I would like to get into a question that I think is really quite important these days, and that's the role women can play in engineering and high-technology development. What motivated you to go into a high-technology area and how has it worked out?

I think the critical element is good training, attending a good school. That means reaching women at the undergraduate and even high school levels. You have to interest them in what's at the other end of all that training. Show them why they should go into technological fields.

Do you think some women might have the motivation but are forced, by certain circumstances, to repress it?

Yes. The image of an engineer isn't terribly glamorous. When you're a teenager, you like glamorous things. Engineers are pictured as fellows in gray suits carrying briefcases and with the ubiquitous calculator hanging off their belt.

You must have had some idea of what science entailed.

Well, I'll tell you, astronomy really grabbed me as an undergraduate. Understanding celestial mechanics was fascinating. Then one thing led to another. Physics turned out to be interesting after all.

Did your interest begin in high school?

No, actually it was at the University of Oklahoma. A spectroscopist named Dr. Betty Pollock received an NSF grant to conduct a physics course for women only. It was a basic physics course that every science or engineering major had to take. She wanted to consolidate the women in one class so that we wouldn't be afraid to raise our hands and ask questions and so forth in class. She also got money from NSF for a very small laboratory facility so that we could have our colleagues within this small group for lab partners. I think there were about 15 of us. We did homework together, and although there was competition on exams, we worked very closely together for two years going through basic physics, physical mechanics, and other classes. We grew very close because we studied together.

It sounds like you had a unique subculture that was quite an advantage.

Oh, very much so. Rather than thinking we were anomalies or bizarre people, we could ask questions in class without feeling intimidated. We grew very close because we had one thing in common—a lot of hard work. If you work on something very difficult with somebody, a bond seems to develop between you.

By talking to your peers in class, you learned the value of different insights; if one didn't have the answer, you looked for another insight.

That's very true.

Was this a four-year program?

It continued for three years. I got in on the second year, but it started when I was a sophomore.

Does it still continue?

No. It was a one-time-only grant from NSF. I don't know what Dr. Pollock ever did with the information. She is a dean at the University of Oklahoma now, I believe. I thought the program was highly successful.

It sounds to me like such programs should be encouraged.

All but one of the women I've kept in touch with are working

> **The image of an engineer isn't terribly glamorous. When you're a teenager, you like glamorous things. Engineers are pictured as fellows in gray suits carrying briefcases, and with the ubiquitous calculator hanging off their belt.**

in the sciences. That one woman has a job as a computer programmer, but she is basically a ballet dancer and uses her computer job to support her ballet.

Are the others all practicing physicists or engineers somewhere?

Physicists, M.D.s, engineers, and geophysicists. One of my best friends from school is now a vice president of an oil exploration company in Oklahoma. Another is in Saudi Arabia doing geophysics and geology. Another is a surgeon at Stanford.

It sounds like a highly motivating course. It's too bad it wasn't continued.

We really learned the subject matter because we were the object of a lot of scrutiny. We dedicated ourselves to do well. In classes which included others than just our core group of 15, we particularly wanted to excel.

It seems similar to the cellular approach to training people.

You read about it in popular management books. The book *In Search of Excellence* discusses task teams. We worked that way at the women's lab; I think it's a very effective approach. You get a group of, say, 10 or 15, to work on a problem. Well, school is just a collection of problems. We all put our heads together and solved the homework problems. Also, physics and engineering people are often loners—I'll lay claim to that one at times too—but if you work in a group, you seem to get the problems solved a lot better.

That's because of the multiple insights.

It sure would be nice to do it in the high school. Forming groups of students to work on scientific projects would probably improve the high school science program.

I would like to talk a little bit about the attitudes of kids in high school. What do you think are some of the things that motivate a person? Not just a woman, but any young person who hasn't thought about sciences but has the capability? Did you take calculus in high school?

No. I didn't take calculus in high school. Scientifically I was a late bloomer. As I remember my high school years, I suppose the teachers made the largest impression, and maybe some outside speakers. I'm trying to influence one of my neighbors in high school who, I think, would be a good engineer. She's reluctant though: "Oh, it would be too difficult. Why should I go through all that bother when I could make more money as a physician?"

Do you think that if people like you were to go into the high schools as speakers, they could be role models?

I think it would help if people would go into the schools and describe their jobs. In optics we have fascinating jobs. When people think of optics, they think of eyeglasses, ergo boring. I think if high schools had more role model situations—seminars, career fairs, or whatever—it would be great.

But you hadn't taken science or physics in high school. What turned you on to science in college?

I had to take a science course for a Bachelor of Science degree—university requirements. Physics was such an intriguing word that I decided to take a physics course and find out what it was.

Did you have any idea what it meant?

I had ideas about Newton's laws and people doing experiments—measuring gravitational constants and things like that. I really wanted to get involved and find out what basic physics was. Simultaneously, I was taking an astronomy course, and really, celestial mechanics and spectroscopy were the most fascinating subjects to me. I found that sociology and other courses at the university were boring compared to the challenge of science courses. My only previous exposure to science was reading *Scientific American*. I used to love to read that magazine.

Do you think that young people who enjoy reading Scientific American *might consider going into science?*

Absolutely. I recommend *Science News* or *Scientific American* as gifts for younger friends to see if they show any interest.

Do you think some people feel somewhat intimidated or overwhelmed by the authority of Scientific American*?*

Oh, perhaps, but I think the younger people I know are so computer oriented that they don't seem to be as intimidated by technology as they would have been a few years ago. My neighbors who are in grade school, junior high, and high school all have computers in their homes. They started out with games, like Pac Man, so they were never afraid of computers. And now they're taking computer courses in high school, and they have computer tapes teaching them geography and other subjects. I don't think that the younger ones are so intimidated by technology. However, I think there's still a big math fear.

Math anxiety.

Math anxiety, thank you very much, that's the term. And I think it's very real.

There's a physics anxiety as well—what to do with an inclined plane.

I think I've had that problem before.

The career path

After you received your PhD, did you examine the range of opportunities before you settled on optical research?

I job hunted for a year before I graduated, and I knew where I was going before I took my final exam.

How did you attack job hunting?

Job interest was number one. What was the organization doing that was going to be interesting? Variety of work was also important.

How did you find that out?

I went to visit the facilities, and coming from a place like the Optical Sciences Center, I often knew people in the companies who had been phasing out of the Center as I was coming in. I had dinner with some of them in relaxed settings so they didn't necessarily have to sell their company and I could ask them honest questions about the variety of their work. Were they stuck on one project, or on projects where funding came and went? Did they have any control over what they did? Was it interesting? Or was it oppressive?

Did you find any reactions to the fact that you're a woman? Or was the fact that you were from the Optical Sciences Center overriding?

Mostly it was overriding. One person was honest enough with me after the job interview to tell me that he thought that age and gender would be a problem because the group that he was seeking a replacement for was made up of people who he didn't think would really be willing to accept somebody with my physical characteristics, if you will. However, I thought it was very reasonable that he said, "I'm going to extend the job offer to you, and I'll look for another section you might be in. But these guys have been together a long time, and I know their attitudes. I think it would be kind of a battle." It was a warning to me.

Were they measuring you on your merits?

I didn't think it went one way or the other, to tell you the truth. A lot of people in school were thinking that women would receive preferential treatment because of quotas. We all compared our job

Nobody wants to be working on a little back burner project that nobody cares about. You like to think that you're in the mainstream of some sort of technological development. I perceived a criticality—that I could make a contribution to the world.

offers, and they were just about the same. I didn't see preferential treatment or discriminatory treatment.

What were the factors that led to your position at the weapons laboratory?

There were three main things that led me to choose the weapons laboratory. The variety of things to do and the autonomy in which you do them were very important to me. Also, the importance of the job as I perceived it. Nobody wants to be working on a little back burner project that nobody cares about. You like to think that you're in the mainstream of some sort of technological development. I perceived a criticality—that I could make a contribution to the world. The third factor was geography. I love Albuquerque. I love the Southwest. I love Tucson. Moving to Albuquerque was one of the best things that we ever did. I just love it.

What type of a role was offered you initially?

First of all, I had responsibility for the optical engineering of a very large, space-based laser weapons demonstration experiment. It was educational for me because I had never seen how very large space systems actually get designed, developed, and put into hardware. Of course, it's a very slow process. I was in on the design stages. Then it was turned over to contractors and to other groups to do the final engineering and hardware development. The project was gradually taken from the hands of the government team, so they no longer needed technical people to do hands-on engineering. The government team needed people who had technical expertise and who could also oversee somebody else's work. I wanted to get out of the role of overseer. It was educational to see someone else's work, but it wasn't really that gratifying after awhile. So I broke off with my own program, and that's what I've been doing for the last two-and-a-half years.

How were you able to establish that?

I went to the director of the laboratory with some ideas, and it was an interactive process. The director then summoned a small group of us and described a technological need. We worked independently on a solution for a couple of weeks and went back. He thought it was a good idea and gave some seed money, and the project and team flourished.

At that time were you the leader?

No, there was somebody with more seniority who was the leader, but as things developed, he didn't want to maintain the lead role. He is a mathematician, and he didn't like being a project leader, per se.

Did the project start off with only a few people?

There were three of us who were all working on the large space program I described earlier. We broke away slowly from our program and started building up our own team, which is now more than 30 people. The group includes 10 women (including the two secretaries). I've also had two great male secretaries.

Are you still working on a particular project, or is this a more general area?

Programs are blossoming again. We have one main project going—phased array telescopes for high-energy laser beam projection. We are taking a number of separate telescopes and controlling them actively to perform as a single beam projector that produces coherent addition of the multiple beams on target. Technical people are never content to grind one application into hardware so we keep thinking of more and more applications. Right now the branch has about five major projects.

Each time you get one of these ideas you grow a little bit.

Yes we do; we work in the task force mode. Somebody in the branch will have an idea and we'll identify, say, three to seven individuals who will probably be interested in it, and they will work on it either full-time or part-time. It simply takes some brainstorming and analysis and maybe something set up in the laboratory.

Does the organization allow you to do this—to sort of seed these things?

Yes. We have some discretionary funds and, after all, the brainstorming is basically free. That is, the cost of brainstorming is civilian salaries. We possess our own computers in the Weapons Lab–a Cray, for example. So we do have a discretionary computer account. The critical element is time. Our laboratory reserves some time for original exploratory technology development. So new efforts are started by individuals who are interested enough in a new idea to take time away from their regular work for a few hours or a few days.

I think there's a funding problem with many organizations. They don't allow the local manager the latitude to do that.

That's really the problem if you're funded by contract funds. In other words, if you work for a corporation where 100 percent of your funding is not internally generated and you're working, say, three programs, then your time card has to reflect that. That leaves you Saturday or Sunday or the evenings. Many of the problems are stimulating enough to do on your own time, but it's significantly better to be able to be creative on the job.

It's very hard to put a team together without a local manager who has the latitude to take time for brainstorming.

It sure is. That's the beauty of either a government laboratory or research facility where you have internal funding and the flexibility to solve the problems.

You're reflecting some of the mechanics of In Search of Excellence. *You were already practicing it without having had it formalized for you. Do you feel that?*

I feel that there is dichotomy in our laboratories. There's a lot of old management structure hierarchy, particularly in the government laboratories—a strict chain of command. Yet when you have creative people pulling together, they seem to transcend the chain of command.

Woman as manager

In these few years you've come from an entry level position with a fresh PhD, and now you are managing a project that involves more than 30 people. During that time you must have observed a variety of management styles. Maybe you could describe some of what you've observed.

Management technique, in my mind, is common sense. And that's really easy to say if you're in on basic R&D. If you have to really produce hardware products, I'm sure it's a lot different. We aren't in the business of grinding metal and putting nuts and bolts together. We have many creative, technical individuals. It's a matter of common sense—finding what people are interested in, what they want to do, what they excel in, and making sure that the problems are there that you can all work on. I really believe in joining in on the problems, too. I think the old idea of the executive manager sitting behind the desk and doling out tasks just doesn't work in this kind of environment. The manager has to be one of the team.

Do you see personality traits that are successful in managerial situations and others that are unsuccessful?

Yes. It's interesting because the dictatorial personality can be successful when there's not enough time to worry about something and mull it over. You need an instant response to something, particularly if it's not in brainstorming phases. So, although that style can work in certain circumstances, it doesn't hold people

> **I think the old idea of the executive manager sitting behind the desk and doling out tasks just doesn't work in this kind of environment. The manager has to be one of the team.**

together. It doesn't make for a longlasting, successful team.

But surely there must be times when you have to come down on someone. The buck stops with you somewhere along the line. How do you deal with that?

Again, mainly by the task force. It's important, before you react, to think about what has to be done. If you have a deadline, an answer you have to provide, a briefing you have to give, a report that has to be finished by a certain time, it's important to figure out in advance when the deadline is, who's got to do what, pull the people together, and keep tabs on a schedule. But, again, if people are working on a team and everybody is doing his or her job, I think the team member who might perhaps solo or procrastinate would be more motivated to meet the deadlines as well.

You might have to make a choice to pull one project off versus another project within your funding constraints and time frame. How do you deal with that?

Yes, it's a tough one. I just tend to try to weed out the programs that aren't doing so well. I wish I could do the same for my stock portfolio! But, at any rate, I try to minimize the programs that look like they're going to be losers and go for the ones that are going to pay off. It's a tough judgment call.

Do you have any observations about women's role in engineering?

It seems that the most important thing, since we seem to be scrutinized frequently, is to make sure that our technical tools are in place and are strong, which results from good technical training. The question seems to disappear when you start working with people. Although there are always comments and questions about whether this is a primary or a secondary career.

How should that question be addressed?

Typically, I don't address it. People who work with me know that my career is very important to me. I guess I'm not much of a person to squabble with meddlers.

You have a positive attitude. But how should upper management handle this? If a woman decides that it is time for her to spend more time at home for a period of 15 years, should management think of it as going to another company?

I just don't see it anymore. It was the case for many years that women would work up to a certain time. But the women with whom I went to school, for example, who had families, haven't quit their full-time jobs. So I guess I'm not in tune with the problem. But, it is still extremely difficult to balance career and family responsi-

I think women are the managers of the future.

bilities. However, the expectation that women will end working careers when they marry and/or have children is diminishing because women are finding ways to get help with domestic responsibilities and child rearing. Furthermore, men are taking a more active role in these aspects of life, and progressive employers are recognizing the value of helping their female workforce meet both professional and family responsibilities.

Maybe women quit those jobs that they want to get out of, but those that are real career opportunities, they don't give up.

Perhaps, but I think many years ago women would quit their jobs because that was the sociological expectation. Now we're seeing the other end of things. Now these women feel they'd like to go back, but it's difficult for them. We see this in our families—our mothers, for example, who quit work to have kids and now would like to go back to work. But now they are running into the age discrimination problem. Personally, I think women in my generation don't want to go through that.

How do women in management roles handle this?

I think women are the managers of the future. I've read in publications like *Megatrends* and *In Search of Excellence* about the necessity of finding the right niches for people. Difficult employees, for example, aren't rotten apples who aren't super achievers; they just don't have the right niche. There's a great deal of truth to that. Of course, there are always some folks who will never be content in any job and will never do good work. Let's push those aside for a second. I think that you can find optimum positions for people if you pay attention to what they are good at, what they're interested in, what they want to do, and then give promotions or awards or some sort of recognition.

Do you think women have that innate ability?

Maybe it's a trained sensitivity. As I said, women are the managers of the future. Still, in order to be a good manager of a technical team today one has to be technically strong, because people in a good team will recognize weaknesses. They don't want some idiot calling the shots.

They must be technically competent.

Yes, and also have an eye for what's important in order to progress as a team. In other words, a good view of what's on the outside that will be bringing in the bread and butter and what's important for the team to be working on.

They need a general outlook.

Yes, they need a good general outlook of whatever world they're working in. They need to know how to organize a team to get things done in an expeditious manner so that working overtime is not a habit. And they must reward individuals with whatever it takes to reward them—promotions, pay raises, plaques on the wall, going to lunch on their birthdays, things like that. I think women are very in tune with that kind of thing.

Brian J. Thompson & James J. Scannell

Pre-college and Undergraduate Education— Challenge for the Future

Background

Brian J. Thompson

James J. Scannell

We are a world economy, more so, with the recent changes in Russia and Eastern Europe. What effect will the liberalization and democratization have on international trade in optical devices and instrumentation, to say nothing about future prospects for technology transfer? What will the emergence of a unified Europe mean for those in the optical industry as we export or import optical goods and services? At home, what will be the effects of a sharp decline in U.S. government spending on the military—both for actual hardware and for research and development? Will some of those funds be transferred to other programs, whether social, environmental, or peaceful scientific and engineering endeavors? What role can and should optical engineering play in the changing patterns of government spending? Can our discipline adapt to changing times and changing support?

How shall we face these issues? The answer lies with education. Currently, the optics industry draws its personnel from a variety of disciplines. There is no centrally focused set of programs to meet the needs of industry. We are not organized in the same way that education is organized in physics, chemistry, mathematics, electrical engineering, etc. Industry draws its optical talent not only from a variety of backgrounds that include the formal optics programs in science and engineering but also, importantly, from a set of scientific and engineering disciplines. In assessing the pool available, we must turn to the total pool of scientists and engineers at the bachelor's level. Richard Atkinson's article in a recent issue of *Science*[1] entitled, "Supply and Demand for Scientists and Engineers: A National Crisis in the Making," makes interesting, if disturbing, reading. He points out, using statistics from the National Science Foundation, that the growth of first professional degrees in science and engineering has had an annual growth rate of 4.8% since 1900, with major dips during World War I and World War II

Brian J. Thompson is Professor of Optics and Provost of the University of Rochester; James J. Scannell is Vice President for Enrollments, Placement, and Alumni Affairs at the University of Rochester.

Thus, as we enter the last decade of the 20th century, we know there are three significant megatrends which all institutions of higher education have wrestled with: the decline in the pool of college-bound high school graduates, the increase in costs, and the declining quality of preparation in primary and secondary education.

and a smaller dip at the time of the Korean War. Periods of rapid increase occurred right after these major conflicts. A slightly higher increase is noted in the early part of the "baby boom" years; however since about 1980 the increase has been essentially zero! And there are more difficult years ahead. At the PhD level of education in the natural sciences and engineering, there was a sharp decline between 1972 and 1980, with a further small decline in subsequent years; now in the last couple of years we have seen that trend reversed. Those statistics hide the fact that an increasing proportion of those degrees are being awarded to foreign students, not all of whom are available for the U.S. workplace. For example, in engineering 30 percent of PhD degrees were awarded to foreign citizens in 1971; in 1988 that percentage had grown to 54 percent. Similarly, in physics and astronomy the percentages were 19 percent and 40 percent respectively.[2]

Challenges for the future

Thus, as we enter the last decade of the 20th century, we know there are three significant megatrends which all institutions of higher education have wrestled with: the decline in the pool of college-bound high school graduates, the increase in costs, and the declining quality of preparation in primary and secondary education. In the case of at least two of these megatrends—and somewhat arguably all three—science and engineering education (that must be thought of as including optics) has been impacted more severely than other disciplines. Not only is the group of college-bound high school graduates smaller but, within that group, fewer students are taking the necessary preparatory courses that would enable them to take advantage of an undergraduate science or engineering curriculum. The number of women attending college has increased proportionately, but their inclination towards a science education has not. Participation rates of Hispanics and Blacks rose through the 70s but declined in the 80s. But, these subpopulations too have shown a disinclination toward the sciences. From 1978 to 1988, the percentage of U.S. Hispanics, Blacks, and American Indians among new PhDs in mathematics, physics, and other physical sciences varied only slightly from one to two percent. All of this is set against a backdrop of need which will place more emphasis on education's relationship to the economy, to productivity, and the health of the nation. At exactly the time when a need for scientists and engineers has never been greater, when the U.S. standard of living is being challenged by the Pacific Rim and Europe, when trade wars are emerging, when performance and results-oriented education is the rule of the day, we are at risk of losing ground and falling far short of producing the necessary scientists and engineers.

University of Rochester—an example

The University of Rochester's College of Engineering and Applied Science is a good case study for what has happened to engineering nationally. It provides some invaluable experiences because of lessons learned with its well-respected, undergraduate optics program and, now, its newly emerging bachelor of arts in engineering program. In the last 20 years, freshman applications to the College of Engineering and Applied Science have more than tripled from less than 400 to more than 1200. Bachelor's degrees have also increased by 300 percent with the high water mark achieved in 1984-85 when 252 degrees were awarded. In 1969-70, only 10 percent of the degrees awarded were in optics (5 of 46). In 1989, 37 percent of the degrees awarded were in optics (67 of 180). Women now comprise 20 percent of the applicant pool to the College of Engineering and Applied Science as well as 20 percent of the graduates with bachelor's degrees in optics. Although enrollment rates for women in higher education have been increasing, in the last five years there has been no increase in their rate of applications to engineering at the University of Rochester. Over the last five years, underrepresented minority student (Black, Hispanic, and Native Americans) applications and enrollment in the freshman classes of the College of Engineering and Applied Science have increased from 5 percent to 10 percent. During that same period, between 5 to 10 percent of the bachelor's degrees conferred in the College of Engineering and Applied Science were received by minority students. Of this latter statistic, however, 84 percent of these degree recipients were Asian, one quarter of whom were students studying on an immigrant visa. Thus, only a very small number of underrepresented minority students received bachelor's degrees from the College during that five-year period.

Conclusion—education systems under scrutiny

We are currently bombarded with articles and studies about education in the United States. We in education at all levels, from preschool to the postdoctoral level, are being scrutinized and criticized. It is important to stress some differences, however. Our education system of K through 12 is being evaluated on its achievements and its ability to change with changing demographics. It has been found wanting by expert after expert. We are falling behind many competitive countries of the world in general education, and particularly in science and mathematics education. We do not need to repeat the facts. Bold measures are needed and programs like the one in the Rochester City School District must be created and then continually evaluated. From our own (science and engineering) parochial view, we must find a way to prepare our high school graduates—and particularly minorities and women—for possible careers in our field. If we do not, our personnel requirements will not be met.

Again, innovative programs such as the Program in Rochester to Interest Students in Science and Mathematics (PRIS^{2}M) need to be in place and supported—similar programs under different acronyms exist in other cities. The PRIS^{2}M program is best described in its own words: "PRIS^{2}M was created in 1978 by the Industrial Management Council, a Rochester business organization. It remains today as a business-sponsored program. PRIS^{2}M's first efforts included programs in two high schools and two junior high schools with large minority populations. Emphasis was on curriculum development, in-service training for teachers in the use of hands-on, problem-solving curricula, the introduction of role models into the classrooms, tutoring, and summer science workshops on a university campus. In 1982, PRIS^{2}M broadened its activity with new programs at 9th, 10th, 11th, and 12th grade levels at all six city comprehensive high schools. The programs are aimed at involving students directly in science-related activities during their high school years when important career decisions are made that affect what they do when they move on. . . . Students with academic potential participate as members of science teams established at the schools. . . . Teams are supervised by teachers, called coaches—one for each high school. The coaches coordinate the efforts in each of the schools and assure support from the educational community. . . . Assisting are PROS*, who are science professionals recruited from the staffs of local industry. These engineers, scientists, computer specialists, and technicians serve as important role models to inspire and motivate the high school students. More than 500 are members of these science teams."

Dr. Chaira Nappi in an article in *Physics Today*[3] makes some interesting comments about precollege science and mathematics education in Europe as contrasted to the United States. She notes that "up to middle school, math and science (especially math) proceed at a much slower pace in the U.S. than in Europe.... A

---

* Professionals

The challenges, therefore, are obvious: better high school preparation in science and mathematics, particularly for women and underrepresented minorities, with increased enrollment and retention of women and underrepresented minorities in engineering and science curricula.

consequence of this approach is that the amount of mathematics that foreign high school students learn over four or five years is concentrated in the last two years of high school in the U.S. These math courses are therefore necessarily very fast-paced and intensive. Moreover, they are usually elective, or optional, courses. It is not surprising that a good 50 percent of American students give up and content themselves with only fulfilling the minimal requirements."

Dr. Nappi also notes that "...girls in European high schools do seem to perform better than their American counterparts. It is not that stereotypes or gender roles do not exist in Europe. They do. However, in more structured educational systems like those in Europe, there is much less room for stereotypes to have an effect." Certainly food for thought.

By comparison, American higher education has not been found wanting in the quality of education it provides. We still are looked at with some envy because of the richness and variety of our programs. If we have been criticized, it is for not controlling the cost of higher education, whether that be cost to the individual, to institutions (endowment draw), or to the taxpayer. We must address the cost issue but, more importantly, we must also not be complacent about our course and program offerings. What was good for the past will certainly not be adequate for the future given that we will have a very different mix of students to educate.

The challenges, therefore, are obvious: better high school preparation in science and mathematics, particularly for women and underrepresented minorities, with increased enrollment and retention of women and underrepresented minorities in engineering and science curricula. Straightforward as the challenges are, solutions are likely to be complex and necessarily comprehensive in nature. Nevertheless, we in education and those involved in science, engineering, and technology have an important role to play. The world is fast changing; our educational system must keep pace.

References:

1. Richard C. Atkinson, "Supply and Demand for Scientists and Engineers: A National Crisis in the Making," *Science*, Vol. 248 (27 April 1990), pp. 413-520.

2. American Institute of Physics, "Human Resources in Science and Technology: Improving U.S. Competitiveness," Commission on Professionals in Science and Technology, March 15-16, 1990.

3. Chaira R. Nappi, "On Mathematics and Science Education in the U.S. and Europe," *Physics Today*, Vol. 43, No. 5 (May 1990), pp. 77-79.

Afterword

A reflection on the beginnings and future of SPIE on its 35th Anniversary

When I came across some literature in the mid to late 60s about SPIE, I wrote a letter to the then new executive director, Joseph Yaver, telling him I thought he was doing things that were very important. I volunteered to get involved. When the need arose for a new editor for *Optical Engineering*, Joe and other society volunteers offered me the editorship and asked me to investigate the need for and role of the journal in the future. It was apparent after many discussions with various people that there was a constituency out there in the field of optics—mainly optical engineering and engineering applied-type optics—that needed to be serviced by a journal. Thus began my long association with the society.

The people who got this society started, who maintained it, and who worked with it through some very trying and difficult years have my greatest respect. They had the foresight and imagination to fulfill a need for the community of photographic engineers and scientists. The society has grown to include many different fields such as optical engineering, optoelectronics, biomedical optics, astronomical instrumentation, and many others, but it was these founding people who were instrumental in the development of the society.

In its 35 years, SPIE has done a lot for education, and continues to do so. It has brought engineers, scientists, and even physicians together, given them a mechanism for sharing information, and has helped to promote the birth and succor of new technologies. It has, by virtue of bringing people together, helped to broaden the scope and outlook of an entire field and created opportunities that might not have been there previously.

I see a very bright future for optical engineering. Technology continues to thrive, and with it, SPIE's role and stature.

North Andover, Massachusetts   *John DeVelis*
June 1990   *Merrimack College*

John DeVelis was editor of *Optical Engineering* from 1973 to 1979. He is coauthor of the book *The New Physical Optics Notebook: Tutorials in Fourier Optics*.

Without a lot of behind the scenes help, this book would not have been possible. Thanks go to Joseph Yaver, Executive Director of SPIE, who conceived and pushed for the idea. In addition: Rich Donnelly, managing editor; Matt Treat, book design; Robin Rawlings, Karen Long, Kate Weisel, graphics and layout; Teresa Larson, Christal Balthazor, transcription; Don Grandstrom, proofreading; Eric Pepper, information and guidance; Lori Fox, permissions clerk; Roy Potter and Rick Feinberg for their technical assistance; Regina Wender, research.